Praise for

The Modern Homesteader's Guide to Keeping Geese

If you are considering bringing geese into your life, this is an excellent
resource to help get you started. In addition to detailed, easy-to-read
information on everything from choosing breeds to keeping geese
healthy, the author's bond with her geese is clear throughout.
Get ready to ditch your pre-conceived notions about geese,
and start enjoying the benefits of adding these useful,
entertaining birds to your backyard or homestead.

— Victoria Redhed Miller, author,
Pure Poultry: Living well with heritage chickens, turkeys and ducks,
and the award-winning book *Craft Distilling*.

It's always a joy to read a book written by someone who
so clearly loves the subject about which they are writing.
Such is the case with this book. Between the beautiful photographs
and charming stories, Kirsten's infectious love of geese has even
made me think about getting a few more goslings for our pond.

— Deborah Niemann, author,
Homegrown and Handmade, Ecothrifty, and *Raising Goats Naturally*
and the blog ThriftyHomesteader.com

Everyone should have geese. Kristen does a great job extolling all
the virtues of these great additions to any farm or homestead.
Read this book for all you need to know about geese...
and then get yourself some!

— Dyan Twining, co-founder Urban Coop Company

THE MODERN
HOMESTEADER'S
GUIDE TO KEEPING GEESE

THE MODERN HOMESTEADER'S GUIDE TO KEEPING GEESE

GEESE

KIRSTEN LIE-NIELSEN

new society
PUBLISHERS

Cover design by Diane McIntosh.
Cover art © iStock.
Background texture AdobesStock_75003007. Feather illustrations © MJ Jessen.

Printed in Canada. Third printing November 2021.

This book is intended to be educational and informative. It is not intended to serve as a guide. The author and publisher disclaim all responsibility for any liability, loss or risk that may be associated with the application of any of the contents of this book.

Inquiries regarding requests to reprint all or part of *The Modern Homesteader's Guide to Keeping Geese* should be addressed to New Society Publishers at the address below. To order directly from the publishers, please call toll-free (North America) 1-800-567-6772, or order online at www.newsociety.com

Any other inquiries can be directed by mail to:
New Society Publishers
P.O. Box 189, Gabriola Island, BC V0R 1X0, Canada
(250) 247-9737

LIBRARY AND ARCHIVES CANADA CATALOGUING IN PUBLICATION

Lie-Nielsen, Kirsten, 1990-, author
The modern homesteader's guide to keeping geese / Kirsten Lie-Nielsen.

Includes bibliographical references and index.
Issued in print and electronic formats.
ISBN 978-0-86571-861-6 (softcover).--ISBN 978-1-55092-654-5 (ebook)

1. Geese. I. Title. II. Title: Geese.

SF505 L54 2017   636.5'98   C2017-904266-1
                                          C2017-904267-X

Funded by the   Financé par le
Government   gouvernement
of Canada   du Canada

Canadä

New Society Publishers' mission is to publish books that contribute in fundamental ways to building an ecologically sustainable and just society, and to do so with the least possible impact on the environment, in a manner that models this vision.

Contents

Foreword

By Lisa Steele

*M*Y EARLY EXPERIENCES WITH GEESE were much like those the author relates at the beginning of this book and usually involved my mother urging me closer to a gaggle of unruly wild geese at the local reservoir or petting zoo so she could get a Polaroid snapshot of me feeding them for our family photo album. I would end up running tearfully into her arms with several hissing, flapping creatures hot on my heels. Later memories would involve stepping over little mounds of goose poop on the putting green of the golf course. All in all, I can't say that geese had ever been on my short list of animals to raise on our farm.

But fast-forward to the present day and happily raising chickens and ducks on a small property in Maine, not too far from where Kirsten lives. She and I met at a Tractor Supply store event several years ago and hit it off. She related to me her challenges keeping her geese out of her neighbors' yards and out of the busy street she lived on near Portland. But she and her partner had just bought a large farm in a very much more rural area of Maine, and she was so excited to expand her flock of geese, along with adding more chickens and ducks, plus goats, to her menagerie.

We continued to keep in touch, and I even managed to pawn off a few unwanted roosters I hatched the following spring on her, since she now had the room to keep multiple males in her expanding flock. She started posting photos of the renovations she was doing on the farmhouse and barn on her social media, and I was enthralled to see her visions come to life. Of course, she also was sharing adorable photos of her animals including baby goats, chickens, ducks, and geese enjoying their new home.

She shared snippets about the geese, and their personalities and loyalty soon became obvious through her daily photos. She pointed out their value in helping to weed the garden and trim the lawn. And I have to say that's what hooked me. The idea of having helpers in the garden that wouldn't scratch up all our plants (like our chickens do) and who actually prefer to eat grass over commercial feed, but who wouldn't eat all our garden toads and "good" bugs in the garden (like the ducks do), made raising a few geese start to look pretty attractive. She even shared with me how if you offer a particular weed to your goslings when they're young, they develop a taste for that weed and will actually seek it out in your yard (buh-bye dandelions!).

Add to that their value as flock guardians, sounding a very vocal alarm if they sense a threat and being a deterrent to many predators just because of their size, introducing a couple of geese to my homestead seemed like a good plan. And once she told me that by hand raising them from youngsters, they imprint and bond with her and are actually not aggressive and mean, but rather endearing and lovable, that was the end of any reservations I had.

Knowing absolutely nothing about raising geese when I first opened Kirsten's book, I feel entirely confident that I could do a good job of it now that I've turned the last page. Her book answered every question I had — and many that I hadn't even thought of. I am delighted to learn that brooding goslings isn't that different from brooding ducklings — which of course I am very familiar with — and that adult geese aren't terribly different from ducks as far as their feed, water, and bedding requirements are concerned.

I hope that anyone picking up this book will seriously consider raising a flock of geese, either alone or in conjunction with chickens or ducks. As for me, I'm excited to begin my journey as a "Mother Goose." Now I just need to convince my husband that we need some geese....

Lisa Steele, author of *Fresh Eggs Daily*, *Duck Eggs Daily*, and *Gardening with Chickens*.

Introduction:
Why Geese Are Right for Your Farm

*A*FTER I PICKED UP MY FIRST PAIR OF GOSLINGS, it didn't take long for me to be head over heels in love with geese. This transformation puzzled many of my friends. Even those who otherwise understood my growing enthusiasm for the farming lifestyle would ask me, "Why do you love geese so much?" The question always made me smile. There are many reasons for my love of geese, some more complicated than others. But the simple and most truthful answer is that, for me, the colorful presence and personalities of geese have proven them to be devoted and reliable companions.

The first geese I ever met belonged to a childhood friend of mine who had a pair of Toulouse geese, birds who loudly honked and fully harassed me. No friendly greeting or warm cuddles there! Yet, in some unexplainable way, they intrigued and charmed me. I was impressed at how my friend was actually comfortable and completely unthreatened by their hissing sounds and warning postures. It became something of a challenge for me to match my friend's confidence and win over her feathered guardians.

Many years later, when my partner and I were first starting our farm, the subject of geese can up, and I leapt at the opportunity. My partner's son asked if geese could join our small flock of chickens. Within a few short weeks, I carried home a peeping box of downy soft goslings and placed them in the new brooder we had set up on our porch.

My relationship with geese was thus cemented. First of all, I recognized how much more intelligent they were than the chickens. I was, and still am, inspired by the way our goslings imprinted on us, their caretakers. Many waterfowl and a few land birds will imprint or bond with their people, but few do so with the gusto of goslings. Their devotion means that, as long as you are around, they are never more than a few feet away, observing and muttering quiet honks. Unlike most creatures, they won't tire of your presence or grow bored.

There are plenty of practical reasons to keep geese. One of the ultimate "permaculture" animals, geese can be used for a wide range of farm tasks. Their abilities easily transfer from one area of the barnyard to another. They can be used for guarding and weeding, their eggs and meat can be eaten and enjoyed, and even their downy feathers are useful. But it is the personality of the goose that wins me over.

Nearly everyone I speak to about geese has a story of being chased, bitten, or otherwise traumatized by a goose. Wild geese will stop at nothing to protect their territory, and domesticated geese retain that fearlessness. Those who remember goose-related traumas are always the people who ask me "Why geese?"

Geese were domesticated over 3,000 years ago, when they were some of the first animals to share their day-to-day lives with people. They were initially raised for meat and sacrifice, but humans quickly learned how useful their loud honking, large eggs, and plentiful feathers could be. By the time of the Romans, geese were a common barnyard animal, and many farmers would selectively breed geese for certain characteristics, such as fast growth or bountiful eggs.

In 500 BC, during the height of the Roman Empire, well-respected Roman farmers were writing about their techniques in goose husbandry and sharing delicious recipes for goose meat. The peasant farmers of the

United Kingdom and Northern Europe were also using geese for their many attributes. When ships sailed to the Americas to start colonies, they certainly brought geese along and introduced them as livestock to the New World. In the early 1990s, a study by the United Nations found geese to be the fastest growing domestic avian species raised for meat.

Geese closely raised around people are not aggressive. They are some of the most affectionate and dedicated animals you can keep. Perhaps part of the charm is knowing you've garnered the loyalty and devotion from such a pugnacious beast, or perhaps, and very likely, they've won *you* over with their quiet coos and clear appreciation for your care and friendship.

Geese are certainly much easier farm animals than many larger livestock, and they require comparatively little upkeep. Apart from chickens, few farm animals will give you such great rewards for such little effort. You don't need to milk or sheer them, they eat very little if offered green pasture, they don't take up much space, and their instincts kick in when it's time to protect, weed, or lay eggs.

That first pair of goslings that I brought home years ago grew to be a handsome and much adored couple. We were distraught when "Mr. Goose" died a few years later, leaving behind his mate who seemed lost for a few days. Then she took to following my partner around, honking with affection, staying close by, never more than a few feet away at all times. She had decided he was her new love. In much the same manner, a group of goslings out enjoying the green grass of spring will jump and come running with their tiny wings flapping at the sound of my voice.

Geese are practical and useful creatures, but we love them for many more reasons than that. Goslings on the farm are a delight and a joy, and the geese they grow into are no less charming. After overcoming an initial fear of geese, many people find that they are some of the most diligent companions you can hope for.

Mr. & Mrs. Goose as brand-new goslings.

Chapter One:

Getting Goslings

THERE ARE MANY REASONS TO CONSIDER ADDING GEESE to your farm: their loyal company and personalities, their protection of your property, the size and flavor of their eggs, their highly prized meat, or their use as effective crop weeders. Whichever of these reasons appeal to you, once you've decided to add geese to your farm or backyard, the first step is to decide where you are going to get your birds.

Geese can be acquired from a number of sources, including local breeders and hatcheries. You can start your search for geese by looking at mail-order catalogs from hatcheries and perusing the internet to find nearby poultry enthusiasts. Even if you do not use hatcheries for your ultimate poultry purchase, their catalogs will still help you understand the various breeds of geese and their different strengths and weaknesses.

Mail-order goslings are eager to get out of their shipping box.

Ordering from a Hatchery

Hatcheries are often the best source for starting a flock of geese, because they will have the greatest selection of breeds. After researching which types you want for your farm, you may find the only way to get your preferred breed of goose is to order through a reputable hatchery.

It is a good idea to look into the hatchery's reputation before ordering, to ensure the birds you get will be healthy and display the proper characteristics of their breed. Internet research is a great way to find out if that hatchery has provided other buyers with healthy goslings. Some breeds are fairly rare, and if you plan to show your geese, they must meet standards set by the American Poultry Association. Many hatcheries' stock doesn't meet those standards, but companies who place importance on their birds' bloodlines will have higher-quality goslings. If you're buying more unusual breeds, remember that good stock is not inexpensive, and companies selling rare breeds for less often will not be selling well-bred birds.

Most hatcheries specialize in mail-order customers and will safely and reliably ship goslings across the country to new homes. The best places to get healthy geese are the ones that focus on waterfowl and have been raising them for generations. Mail-ordered goslings will arrive peeping via an overnight shipment, in a breathable box lined with shavings. Your post office will call you when they arrive. You need to be ready to pick them up when the call comes, as it is important to get your goslings into their brooder quickly so they can start eating and drinking.

If you are lucky, a reputable hatchery may be located near you, and you can avoid the mail-order option.

Finding a Local Breeder

Some of the healthiest, most robust goslings can come from a local breeder who has spent time and effort making sure their birds are of the finest quality. Good-quality show birds have strong bloodlines that ensure they will display the proper characteristics of their breed. It can be hard to track down a good goose breeder, but it is worthwhile if you plan to show your birds or you are looking for perfect examples of a specific breed.

The average backyard farmer will have mixed breeds that may not display the characteristics you are looking for, but these birds can still make exceptional pets or guardians. If showing or breeding your geese is not in your long-term plan, mixed breed goslings may be just what you are looking for.

Geese from breeders, unless picked up in their first day or two of life, may have already imprinted on their mother or the breeder, and they may not be as friendly as a goose who has only ever known you since hatching. If your geese are being used for guarding and weeding, that bond may not be as important to you. However, if you want a family pet, make it a priority to get your goslings as young as possible so that they will imprint on you. To that end, when getting goslings from a local farm, you can sometimes establish a relationship with the farmer before your geese arrive and be ready to pick them up as soon as they hatch. A relationship like that can also be helpful with your goose-keeping later on, as an experienced goose owner can help you with questions you have throughout your goose's life.

Goslings ordered from hatcheries generally ship on the same day they

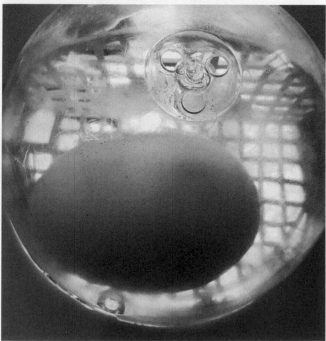

hatch, and arrive to you within their first two days of life. This greatly increases the chance that they will imprint on you, and if you're picking up birds from a local farm or breeder, the earlier you can pick up your goslings the more likely they are to bond with you.

Incubating Hatching Eggs

If you have an incubator, you can try hatching goslings on your own. Goose eggs require careful turning and temperature monitoring, so there is a time commitment involved, but a gosling hatched at home will be healthier than one that has shipped across the country and will almost definitely imprint right away. Like goslings, hatching goose eggs for home incubation can be ordered online or picked up at a local farm.

An incubated goose egg takes 28 days to hatch. For the first 25 days of incubation, they should be kept at 99.5° Fahrenheit, and turned 3 to 7 times per day. The last 3 days, the temperature in the incubator can be lowered to 98.5°, and you can stop turning the eggs. The humidity in the incubator should be 86% for the first 25 days, and 94% for the last 3. The best way to maintain high humidity is to regularly mist the eggs with a household spray bottle, or leave a small tray of water in the incubator. Most home incubators come with a humidity monitor and thermometer, but if yours does not, you can pick up one at your local pet store where they sell them for reptile's terrariums.

Hatching goslings at home is one of the most sure-fire ways to have them imprint on you. Yours will be the first face they ever see, and you'll be the first person to feed them. From that first day they hatch, they will think of you as "mother goose" and should be very dedicated pets.

Local Feed or Garden Supply Stores

Local feed or garden stores may sell goslings in the spring. In February or March, they will often have order forms available at the checkout for deliveries of birds expected to arrive that spring. Even larger chain stores such as Tractor Supply frequently have goslings on their order forms. The choices for breeds might be limited, and the birds' health can vary depending on the source, but if you develop a good relationship with the staff, you may benefit by making connections that can offer you valuable advice throughout your goose's life.

Goslings develop the famous goose attitude early on.

Getting Adult Geese

Of course, starting with goslings is not the only option for introducing geese to your farm. You can also find adult birds from local farms. If you want to start reaping the benefits of keeping geese right away, having an adult goose might be a worthwhile beginning, especially if you don't have the time or energy that caring for goslings requires. While an adult goose will never imprint on you the way a gosling does, you can find birds that are very friendly. Moreover, mature geese are already practiced egg layers and more than likely prepared to protect your property as soon as you bring them home.

Hatching with a Broody Goose

If you already have some adult geese, you may decide to allow one of your females to sit on eggs in the spring. You can purchase hatching eggs, and instead of incubating them yourself, and worrying about humidity and turning times, you can allow the mother goose to take care of the business of raising goslings. This takes a lot of the guesswork out of raising geese, as a mother goose usually knows what to do without needing to do any research.

If you have a pair of geese, male and female, you may also decide to let them hatch their own eggs. Some breeds, specifically Sebastopols and Dewlap Toulouse, have issues with low fertility, and if you wish to breed these, it's important to start with strong stock and allow ratios of one gander to two or three geese. Other breeds have no such issues, and with no effort on your part, they can start a family.

While nine out of ten hatchings go without issue, it is a good idea to calculate when the eggs will be due (about 28 days after your goose starts setting), just in case the mother is too rough with her hatchlings. It is important to give her a separate area that the rest of the flock can't reach, because they will harass her and potentially harm her goslings when they are too young to defend themselves.

Many geese make excellent, protective parents.

Raising Lady Goose

Our first pair of goslings arrived on a cold, wet day in late June. The post office called us to let us know that our African goslings were ready to be picked up, and we jumped in the car and hurried down to get them. The goslings would spend their first few weeks in a make-shift brooder box in our mudroom, and at every passing person, they would jump up and down, peeping to be recognized and played with. Their bond with us required no work on our part, as they instantly imprinted on my partner and me.

A few years later, when one of the geese was taken by an eagle, the survivor, Lady Goose, became head over heels in love with my partner. She spent that entire fall only one or two feet away from him, honking softly in his ear while he worked on our vintage tractor, or calling out urgently for him when he was out of sight in our home. After we expanded our flock, Lady Goose found a new goose companion, and while she remains attached to us, she's no longer as underfoot as before.

Goslings hatched under a mother goose will almost never imprint on a person, because they have their real mama right there to take care of them. If the mother was hand raised and is friendly, however, you'll probably end up with friendly and unaggressive young geese from her.

Goslings and Brooders

Now if you've decided on beginning with goslings, and you know where you'll get them, it is time to get the brooder ready! Goslings are much hardier than baby chicks, but they still have special requirements to ensure they will grow up into healthy, happy birds.

Lighting and Temperature Control

Day-old goslings need a brooder temperature between 90° and 95°F. The easiest way to accomplish that warmth is to use one or two regular 60 or 100 watt light bulbs. You can also use special purpose heat lamps; however, the risk of fire is much greater with these lights. Regular light bulbs will keep an enclosed space warm, and as long as the light is not coming into contact with hay or shavings, the fire risk should be minimal.

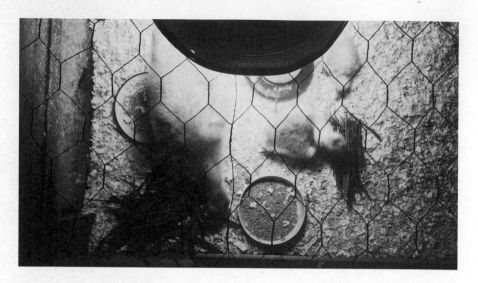

You can reduce the temperature in your brooder by about 10 degrees per week until it is the same as the room temperature, by removing one of the lights or continually raising the height of the light source. In order to make sure that your goslings have a comfortable temperature throughout their growth period, put a thermometer in the brooder and monitor its progress.

If goslings are panting with their beaks open and not huddling up together, they may be too hot. Similarly, if they are constantly in a tight huddle, you might want to increase the temperature in the brooder a few degrees. Keep a close eye on your new baby birds in their first few days of life, and you will quickly learn how they behave when they are happy and comfortable.

Food and Water

Goslings need constant access to food and water. When they first arrive, they may have difficulty eating or display a lack of interest in dry crumble, so it is a good idea to soak their food in water before offering it. Wait until you have a soupy consistency, and your goslings will be sure to gobble up the crumble. Never give goslings medicated feed or chick starter, which is not formulated for them.

In order to eat, geese need to be able to submerge their beaks fully in water; otherwise, their nostrils can become plugged with remnants of food, preventing them from breathing. A classic chick waterer with a plastic trough, which is available in any feed store, allows them to dip their beaks fully underwater. A homemade alternative is an empty 5-gallon milk jug with small holes cut into all four sides, large enough for your goslings to fit their heads through comfortably. Fill the gallon jug up to the holes with water, and this will allow them to douse their beaks without creating a mess.

Goslings should not be offered swimming water at first. Their downy feathers do not have the waterproofing that adult goose feathers have, and they can easily catch a chill if they get soaked. Wild goslings are able to dry off in their mother's downy wings after a swim, but goslings in a brooder can lose body heat quickly.

After a week or two, you can start introducing the young birds to supervised swimming. Just remember to towel dry them gently after they bathe. When their pinfeathers start coming in, they should be able to swim regularly, as long as the water is shallow and easy for them to enter and exit without assistance.

Keeping the Brooder Clean

Pine shavings, hay, peat moss, and straw are all suitable bedding options for young geese. Although I use straw for my adult geese, I recommend a thick layer of pine shavings in the brooder because they are highly absorbent and won't hold

Goslings' first swim!

an odor, which is very helpful with goslings' messy droppings and poor table manners. The most absorbent bedding is peat moss, but is often exorbitantly expensive. Pine shavings and straw are both available at garden centers and feed stores in large bales.

Geese are naturally messy creatures. Even without open water, there will be spills that soak through the shavings quickly, meaning regular changing of their bedding is required to keep the brooder smelling fresh and to prevent possible diseases. Be prepared to do a quick clean and refresher at least once or twice a week — more often as your geese get larger. Once they are 4 to 6 weeks old, it will be time to start transitioning them to their adult housing.

Treats and Outside Time

Because they are hardier than chicks, goslings can start exploring the outdoors earlier. With your close supervision to keep them from getting lost

*Young goslings enjoy a
dandelion treat.*

or attacked by a predator, baby geese can enjoy an hour or two out on the
grass within their first few days of life. They will love being able to stretch
their legs outside the brooder, and relish the opportunity to munch on
fresh grass stems. Taking the time to sit with them during their outside
activity also helps to ensure that they will imprint on you.

Unlike ducks and chickens, geese are vegetarians. Some geese will nibble up mealworms, but their favorite treats will always be fresh greens.
Keeping a good supply of green leaves in their brooder is a great way to
ensure they get the nutrition needed to grow up healthy and strong. You
can offer them fresh-cut lettuce or spinach from your garden, or even regular grass clippings. My geese have shown a particular proclivity for Swiss
chard and beet greens. You can even try hanging a head of lettuce from
the top of the brooder for them to nibble on as they desire. Plants rich in
vitamins and minerals, like beet greens or kale, are also especially healthy
for your goslings to enjoy.

Sexing Goslings and Adult Geese

Some hatcheries will allow you to order sexed goslings, and many will even band male and female baby birds' legs with different colors. However, most of the time, sexed goslings are only available at a higher price, and some hatcheries do not offer this service at all. Of course, if you hatch goslings from an incubator or under a mother goose, you're going to have to determine the boys from the girls yourself.

Even with adult geese, it's often almost impossible to tell males from females except during the mating season. Males are often a little bit bigger, their voices are higher, and they're more vocal. Females are usually more docile in their behavior. Since two geese of the same sex will often pair up if there are no other geese around, you can only be certain you have a male and a female when one starts to lay eggs.

You can vent sex a gosling as early as one day old, and the principles remain the same if you are sexing adult birds. It is not always easy to tell if you do not know what you are looking for, but you can try vent sexing by placing the bird on its back and using a gloved and lubricated finger to open the vent beneath the bird's tail. When gently massaged, the male's sphincter muscle relaxes and his corkscrew shaped penis will pop out. While this is possible to do with adult birds as well as goslings, it is usually easier to handle baby goslings than larger, fully grown birds.

End of Life

Before bringing geese into your life, it is important to consider how you will care for them as they get older. Geese are not typical pets; they can't be kept in a cage all the time, and they will outlive many other farmyard animals. Geese can live between 15 and 20 years, with the oldest known goose dying at 49. This means that they will be part of your life for a long time, and because they imprint so closely with their caregivers, they will be as devoted to you as the family dog. If your lifestyle is transient or city-bound, geese may not be the beasts for you. But if you're looking for a companion to roam your yard or farm for a decade or two, a goose is the perfect choice.

Many farmers raise geese for food, and their dark, gamey meat is a holiday delicacy in many homes. If you have decided to raise goslings to eat them, keep this goal in mind as they are growing up. I have spoken to countless goose keepers who originally intended to eat their birds, but ended up becoming too attached to send their beloved pets to the butcher. Remember your goals with geese and try not to become too attached to them if they are headed for "freezer camp" in the winter.

If you are keeping geese for meat, they will meet their end at your hands or the hands of your butcher in the fall or the following spring, when they are about 15 or 20 weeks old. If your geese are farm guardians, weeders, or pets, and they escape predators throughout their life, they will live on your farm for 15 or 20 years. When they become infirm, you can arrange to have them butchered or work to make their area as comfortable as possible for their last few years. Like many birds, geese will often disguise signs of discomfort, and you may be surprised to find yours in sudden failing health after many years. Geese are private animals and, like an old dog, may wander off to a secluded spot when they feel their time has come.

Raising goslings can be a time commitment, but the close relationship you can develop with young geese will reward the patient farmer. Goslings make a refreshing addition to the farm, and raised with care, they will grow up into friendly, curious companions.

Chapter Two:

Breeds of Geese

$\mathscr{L}$IKE CATS, DOGS, HORSES, AND CHICKENS, GEESE come in many different shapes, sizes, and temperaments. While a goose's behavior can be greatly shaped by how it is raised, each breed has distinct traits. Selecting the right breed for your farm will help you be sure you'll get the most from your bird. Some breeds are loud, excellent guardians and not suited to an urban farm. Others are docile, perfect for families with children, and some are showy, with crowd-pleasing feather patterns.

You can get farmyard mix geese from local farms, which have uncertain heritage and are often all gray or white and gray. Mixed stock farmyard birds like this can make the perfect family pet, but if you have any intentions of breeding or showing geese, it is important to start with strong, purebred stock. After researching your options, select breeds that will suit your type of farm and are predisposed to behavior you think you can handle.

African and Chinese Geese
History

Most domesticated geese trace their lineage back to a common ancestor, the Greylag Goose (*Anser anser*), but this is not the case for the distinctive African and Chinese geese. These birds have Asiatic heritage and descend from the wild Swan Goose (*Anser cygnoides*), a breed still found in northern China and southern Russia. The actual continental origins of these two geographically named breeds are uncertain, but it is thought that they are called African and Chinese simply because they were introduced to Europe by those who had traveled to exotic locales.

A trio of African geese.

Elegant and graceful, both the African and Chinese geese were introduced to the American Poultry Association's Standard of Perfection in 1874 (source: Livestock Conservancy). Since that time, different color variations have emerged, and the apparent distinctions between the two breeds have grown more clear. Always popular on the homestead, both breeds are known for their versatility as egg layers and meat birds, but they are most prized for their vocality and weeding abilities.

Appearance

What's the difference between African and Chinese geese? Their size. Africans are bigger, weighing up to 20 pounds, while the more petite Chinese average only 10 pounds. Both are most commonly found in the brown variety, with black bills, pink feet, and a cream underbelly. White Chinese are solid white birds with orange bills and feet. For that matter, there is a white variety of the African, but these are highly unusual.

The most distinctive feature of these Asiatic geese is the large knob at the top of their beaks. This fleshy bulb is solid and proud, and while its evolution is unknown, a prominent knob is now a sign of well-bred African or Chinese geese. The only disadvantage of this remarkable characteristic is that it can be prone to frostbite.

Both birds have a pronounced upright carriage and long, slender necks. African geese should have a fleshy dewlap under their chin and a noticeably full abdomen. They are keelless, having no large deposit of flesh in the chest area, and carry their tails in a rigid, upright manner. Chinese geese are more slender, with longer necks that make them perfect for weeding. With erect tails, no keel, and a tighter abdomen than their heavier cousin, they have the lithe appearance of well-conditioned athletes.

African goose in the foreground.

Temperament

Chinese geese are the variety most commonly used for weeding tasks. They are high-energy, making them excellent foragers, and their long necks are able to reach in and pluck their favorite weeds from hard to reach spaces.

Both African and Chinese geese are known to be quite vocal. They make excellent guard animals, sounding an alarm at any new activity on the farm, and their piercing vocalizations can be heard across the barnyard. They also can tend towards aggression, though hand-raised birds remain friendly. African geese can be more laid back, but they are still known for their loud voices.

Sebastopol Geese

History

There is not another goose with more distinct feathering than the Sebastopol. Its ragged, unkempt looks could not be mistaken for any other breed, and it is certain they created quite a stir when first introduced to European poultry fanciers in 1860. While the exact origins of the Sebastopol, also

A pair of Sebastopol ganders.

called the Danubian, are unknown, it is thought to have originated in the Black Sea region. Known in German as Lockengans or "curl gooses," the Sebastopol has remained a popular exhibit bird that continues to impress farmers around the world.

Appearance

The feathers of Sebastopol geese have unique soft quills that allow them to bend and curl. Their chest and back feathers are soft and light. These small birds only grow up to around 10 pounds. Their bright blue eyes are deep and thoughtful, and their beaks and feet are bright orange. Their loose, cloud-like feathers toss about in the breeze. Sebastopols are not as cold hardy as other geese, requiring a warm space out of the wind to avoid a chill.

Sebastopol goose.

Temperament

Sebastopols are soft-spoken geese, not prone to honking when it isn't necessary. They are shy and pleasant, and rarely aggressive. As average egg layers, female Sebastopols can lay 20 to 30 eggs in a season, but these birds struggle with low fertility rates. When provided with plenty of deep swimming water in which to breed, a pair can be successful at hatching a clutch of fuzzy offspring.

Thanks to their kind temperaments and spectacular feathers, Sebastopols are a family farm favorite. They are gentle with children, especially when raised by hand, relatively quiet, and sure to draw the neighbor's compliments with their remarkable feathers.

Pilgrim Geese

History

Pilgrim geese, which likely descended from a variety of farmyard mix goslings, are prized thanks to their unusual auto-sexing feathers. Auto-sexing is a quality almost unique to Pilgrim geese (a few European breeds also have this feature). This means that the different sexes can be easily distinguished by the color of their feathers.

While Pilgrim geese get their name from a myth that they were brought to America by the first European settlers, in fact this goose was cultivated by Midwestern farmers during the Great Depression and finally won admittance to the American Poultry Association's Standard of Perfection in 1939. Oscar Grow, a waterfowl breeder in the 1930s, was the first established breeder of Pilgrim geese and named them for his wife's family's frontier journey.

Appearance

Male and female Pilgrim geese have singularly different feathers. Male goslings are yellow with some gray tinting and pink beaks and feet and grow into white or piebald adults with orange beaks and feathers. Females are hatched mostly gray with gray beaks and feathers, and as they mature, become almost solid gray with some white feathering on their underbellies.

Pilgrims are a medium-weight goose and solid egg layers, one of the reasons they were so prized by Depression-era farmers. A goose could be kept for a few years for eggs, and then butchered for the family dinner table.

Temperament

Pilgrims are even-tempered, not prone to big displays of aggression. Males can be protective, so it is important to hand raise them if you want a docile bird. They are not loud individuals, but a flock can cause quite a cacophony. Developed out of a need for a goose that could do everything for the destitute farmers of the 1920s and 30s, Pilgrims are versatile backyard birds and sweet-tempered enough to serve as family pets.

Embden Geese

History

Developed in the early 1800s, Embden geese trace their heritage and name to Northern Germany. By 1820, the breed was imported to America and soon admitted to the Standard of Perfection in 1874. A large, full-bodied goose, Embdens were bred primarily for the dinner table and prized for their fast rate of growth. Above average egg layers, they became popular alongside Pilgrim geese as an all-around bird perfect for the family farm.

Appearance

Embden geese stand tall with rounded bodies and full abdomens, their girth accentuated by their short necks, legs, and tails. Solid-bodied, displaying snowy-white feathering with orange beaks and feet, they have steely blue eyes. They are the second-largest breed, behind only the Dewlap Toulouse; adult males can weigh up to 30 pounds.

Temperament

Like many larger breeds, Embdens are fairly laid-back and even-tempered. They are not known for their loud voices, but they are renowned as excellent egg layers and will often go broody to raise their own clutch of goslings. Unlike some breeds, Embdens have no trouble with fertility, and this combined with their parenting instincts makes them one of the more prolific breeds on the modern American farm.

Toulouse, Dewlap, and Production Geese

History

Toulouse geese come in two distinct types, a common production variety and a more unusual, heavy-weight breed known as the Dewlap Toulouse. The origins of the latter were most likely of one moderate-weight goose, but as some farmers started to develop heavy birds, the distinction between production and Dewlap became clear.

While production Toulouses have been excellent utility birds and a very common sight on farms for the past century, Dewlaps are much less

Two-year-old Dewlap
Toulouse goose.

common and singular in appearance. They were developed out of a desire for heavier and heartier birds. Before they became synonymous with *foie gras*, the French delicacy, Dewlaps were already prized for their high fat content at a time when goose fat was valued as a general lubricant. While some farmers raising Dewlap Toulouses specifically for foie gras keep them caged and force-feed them large quantities of grain, when given access to open pasture, these geese will still put on a great amount of weight quickly, making them an ideal table bird.

Appearance

Production and Dewlap Toulouses have similar gray feathers, with white undersides and orange legs and bills. The former have elegantly curved, smooth necks and tight abdomens, making them easily distinguishable from the Dewlap.

Dewlap Toulouses, meanwhile, could never be confused with any other breed of goose. With a deep, full abdomen, they will develop full lobes (the fleshy area between the legs) that often drag on the ground. Their keels are heavy and round. Adults develop their namesake dewlap, a full paunch of heavy skin hanging beneath their stately beaks.

Adult Dewlap Toulouses can weigh upwards of 30 pounds, while production birds rarely top 20. Both are moderate egg layers, but Dewlaps lay eggs to match their massive girth, often considerably larger than other goose eggs.

Temperament

Production Toulouses have become so common in farmyards that they are almost always a feature of a farmyard mix gosling. These birds can be aggressive if not handled regularly. Because of their well-established presence across America, it is easy to find them or a bird that traces its heritage to them.

Dewlap Toulouses, however, seem to feel that food is more important than attitude. These mostly gentle giants have a placid disposition and will always be the first ones at the food trough when grain is offered. While their size and rarity might indicate that they are an exclusively special-purpose bird for meat production, in my experience Dewlap Toulouse geese make excellent pets. They are not overly vocal; though when they do honk, their call can be deafening. Their Mother Goose appearance is charming, but it is their laissez-faire temperament that will win over any farmer.

Buff Geese

History

American Buff geese have feather colors not found in any other North American breed. Developed from the Brecon Buff, a British breed still popular in the UK, they were refined in the United States and admitted to the American Poultry Association's Standard of Perfection in 1947. Since then, they have remained relatively rare, but an enthusiastic favorite of the farmers who raise them.

A young Buff gander.

Appearance

The apricot-colored feathers of the Buffs give them their name. Medium-weight, with deep brown eyes, orange bills and feet, they have a chunky body with a full abdomen. Their necks are short but arched, giving them a pleasantly swanlike appearance as they glide across the yard. They are decent layers and prized as exhibition birds, thanks to those fawn-like feathers.

Temperament

Buffs, not much for shows of aggression, are sometimes described as curious; they will nibble around a stranger's feet in a display that can seem intimidating. But they are friendly with the people who raise them and are mostly calm, quiet birds. Buffs can go broody easily and make excellent parents, though reproducing their striking colors requires careful selection of breeding stock.

Roman Geese

History

Perhaps the oldest breed of goose still being farmed today, Roman geese trace their lineage back to the Great Roman Empire. It's said that these small birds protected the Temple of Juno during an attack by the Gauls in 390 BC. Since then they have remained great guarding geese. Their stature make them easy to raise on a smaller farm.

One-year-old Roman Tufted gander.

Appearance

Roman geese come in classic and tufted varieties. Classic Romans are solid white, with orange beaks and feet, and a small plump body. Tufted Romans have the same stocky, full-bodied appearance, but their feathers point upright in a crown-like tuft at the crest of their head.

Romans are one of the smallest type of geese, adult males weighing in at 10 pounds or less. Though their necks are short, they are elegantly curved, and their eyes are a deep, stunning blue. Their small bodies are tasty eating, and they produce an average number of 20 to 30 eggs every spring.

Temperament

Despite being vocal birds and excellent watchdogs, Roman geese are not overly aggressive. They tend to be curious and become easily attached to the people who raise them. You can find a wide variety when it comes to the quality of stock of Roman geese. Look for well-behaved birds with straight tails and prominent tufts, recommended attributes if you want show-quality birds.

Unusual Breeds

Many more breeds of geese can be found on small farms across America. Each has its own special quality and personality. Some hard-to-find breeds in the United States are well-known in Europe.

Cotton Patch geese, like Pilgrims, are sex-linking geese. Adults often have more mottled feathers than Pilgrims, with a slender build and upright stance more like that of a wild goose. Cotton Patch geese, unlike most domesticated breeds, are excellent fliers, thanks in part to their light weight.

Pomeranian geese are remarkable piebald birds with white and gray feathering. Originally from Germany, they are a very ancient type of medium-weight goose. Ideally marked Pomeranian geese have dark heads, white necks and cheeks, dark wings, and white under their tails.

Steinbacher geese are very rare in the United States, but can be found on European farms, especially in Germany. Not a typical gray goose, they are remarkable lavender-steel color. These elegant birds are about 15 to 20 pounds and have long, slender necks. Steinbachers are especially calm and docile, making them wonderful pets.

There are even more unusual breeds of geese to be found traveling the world, such as the West of England goose, a piebald sex-linked bird; the Czech goose, a very small and stocky white goose; and the Egyptian goose, whose rainbow feathers are unmatched. Some intrepid farmers have even cultivated the iconic Canadian goose, though their flight feathers must be trimmed to keep them from migrating.

Your farm's needs can be matched by many goose breeds, but the perfect one will fit into your homestead's rhythms without a catch. For example, the Chinese goose is not a good match for the urban farm, but weeding several acres of orchards could be its ideal home. The Dewlap or the Sebastopol make wonderful pets. And in between, such versatile utility birds as the Pilgrim and the Embden are ready to offer your farm fresh food and plenty of charming attitude. When you choose the breed with the characteristics you're looking for, you can find years of satisfying companionship and service.

Chapter Three:

Feeding and Housing

$\mathcal{C}$OMPARED TO MANY LARGER LIVESTOCK ANIMALS, geese are easy keepers. Of course, they do require shelter and nourishment, so you'll need to provide proper housing, food, and plenty of water in order to keep your birds happy and healthy. Staying attentive to their basic needs will help ensure your flock is a productive part of your farm.

Young Geese

Young goslings can be started in a brooder box similar to ones used for baby chickens, ducks, and other poultry. Four to six goslings can be kept comfortably in a brooder that is two feet by four feet until they are about four weeks old. The space you set up for your goslings should be kept at the appropriate temperature with a light or heat lamp and should have tall sides (at least two feet high) to prevent them from jumping out. A chicken wire lid will keep goslings from escaping and prevent curious pets from bothering them. The brooder floor can be lined with your preferred bedding: pine shavings, straw, or peat moss. It needs regular cleaning, as goslings saturate their bedding with droppings and spilled water.

A trio of goslings, about three weeks old.

At three or four weeks old, your young geese will start to feel confined by a brooder box. It is best if you have a temporary grow-out area for them. If you already have adult geese, try to provide an area for goslings that is visible to your adults without allowing them to physically interact. A fenced-off area within the adult living quarters is perfect, since that arrangement allows both age groups to get used to each other without being aggressive, pecking, or showing other signs of harassment. If you do not have adult geese, you can move your goslings directly into their adult quarters, as long as the area is secure and warm.

A stall inside a barn, shed, or outbuilding may be perfect for housing your geese. If you're planning to keep them with other birds, then this age is the perfect time to start introducing them. Fencing off a corner of a chicken coop for goslings allows both parties to get acquainted before they start living together full time. If you are housing geese with smaller birds, it is important to make sure a part of their coop is accessible only to the smaller birds, in case your geese bully them.

At night young geese need a secure, warm space for sleeping. An area allowing about one-half a square feet per bird is perfect. You can still provide a light for heat at this age, so they can gather under it when they are chilly. Many farmers shelter their geese in specially built houses, similar to a dog house or a duck run. A building like this is satisfactory provided it has a door that can be secured at night. This allows for

This pair has already outgrown their brooder box.

At a few weeks old, goslings love to get out and enjoy fresh grass.

easy pasturing, as you can fence the area around the coop and just open the door to allow them out in the mornings.

Immature geese should not be allowed outside unsupervised simply because they are not large enough to protect themselves against predators. They can spend their days in a fenced area near your barn or shelter, as long as you are close by to attend to them. An outside pen should allow them room to roam, with access to water and food and a shaded area for them to escape the sun. At 4 to 6 weeks, they can also try out their swimming skills. If you have a pond or kiddie pool for them to wade in, that will provide a great start. Again, keep a close eye on your goslings and make sure they can enter and exit their swimming area comfortably by themselves.

Adult Geese

Adult geese need more space than young birds, but rarely if ever need supplemental heat. A barn stall or special coop is perfect for your grown flock, as long as it allows four square feet per bird. It is important that their home is dry, protected from blowing winds, and provides good ventilation.

Geese don't mind snow as long as the temperature isn't too cold.

Water spillage and goose droppings tend to build up condensation in winter, which can cause respiratory issues for birds. A vent in the top of your coop or small window will allow airflow without letting cold breezes blow through.

Besides cold condensation, the biggest environmental threat to geese is freezing, blowing winds. Loose-feathered breeds such as Sebastopols are especially vulnerable to cold winds. Make sure the shelter walls are solid and that the birds always have the ability to get in out of the wind on a cold day. In addition, the outdoor area should have some shade for the warmer summer days when geese need to get out of the direct sun, although they generally tolerate heat as long as they have plenty of drinking water. As with the cold, certain breeds like the Dewlap Toulouse tolerate heat less than others.

A door and ramp for geese housed in a traditional New England barn.

A barn is highly suitable for housing geese, since most have plenty of ventilation and provide adequate protection from wind and weather. Geese can be kept happily in a barn stall as long as the main doors of the structure are secured from predators each night. You might provide a smaller, bird-sized door, dedicated to their area. A private goose-entry can be very simply designed; all that is necessary is an opening that allows easy access to the outdoors and back inside whenever they feel the need for shelter or to visit their nests. The main idea is to have ventilation, yet provide protection from the weather and predators. If the barn's ground floor is elevated, a ramp can easily be added, giving your birds the freedom to go in and out whenever they please.

The outdoor pasture area for your geese might be a free-range space, or a fenced zone along the side of the barn. Their ability to open the door into this pasture will save you, the goose keeper, time and effort herding them in and out each day. Of course, you'll need to latch the door in the evening, when the birds are in for the night.

A specially built coop is another good option for housing geese. Like a dog house or duck run, it can be free-standing, as long as it provides the needed footage for them to be comfortable, four square feet of floor space per bird. The roof should be at least five feet high, but you might consider going taller to provide better ventilation. Moreover, it is considerably easier to clean in a full-sized building where you can stand upright to rake out debris. A free-standing goose coop, again, can have a doorway for your geese that is kept open during the days, permitting passage in and out of free-range space or a special pasture. If you already have a chicken coop, your geese can be kept in this space as long as it is large enough and your chickens have

safe roosting space. This is often the best option if you have only one or two geese, since they won't require a lot of floor space.

If you're using chicken tractors to move your other birds around, a similar setup can work for geese. Chicken tractors are coops on wheels that can be dragged from one spot to another, often with a small fenced area for the birds to pasture in. They allow the birds to feed on new areas, either fenced or open, every day, since the housing is easily transportable. The number of geese will determine if such a system is appropriate. If you only have one goose and a number of chickens, then you might want to keep the goose with your hens inside a tractor for the smaller birds' protection from predators. But a moving goose house is not usually practical for a large flock of geese. A dozen or more birds used for weeding, for example, are probably better off in more permanent housing, since it would be easier to herd them out to pasture and back in for nightly shelter than wheeling a larger, more cumbersome tractor unit.

Bedding

A goose house can be bedded with straw, shavings, sawdust, or peat moss. Consider that geese are large birds that leave notable droppings, and they can also be very messy around their water. Regular cleaning of the area keeps it free of odor and disease. You do not need to completely empty their house down to the floorboards when cleaning, but do remove any wet or soiled litter and replace it with a fresh layer of bedding every week or two.

Some farmers use sawdust, newspaper, or sand for bedding. The two most important factors in selecting bedding are the absorbency of the material and the cost. Peat moss is the most absorbent of the above options, but it is often prohibitively expensive. Straw and hay are not particularly absorbent, but they can be relatively inexpensive, depending on your area and the time of year. You can often work a deal with a local store to get their unsold newspapers, but these generally just won't do the job of soaking up the geese's mess. Sawdust and shavings are relatively absorbent and often affordable. This makes them the best of both worlds, but the downside is that they can be fairly dusty. Sawdust, especially, can leave a fine film

Straw makes excellent bedding for geese as well as chickens.

of dust, but this may be worth it for its ease of cleaning and limited effect on your wallet.

Flooring

It might not seem like that important of a choice, and you may be limited in the changes you can make to existing buildings, but it is good to think about the type of flooring for your goose house. If it is built on concrete or free standing on dirt, it is extra important to provide deep, clean bedding in the winter. Both surfaces become very cold when the temperatures drop, and extra bedding will keep the coop cozy. Dirt floors can be hard to clean, as your shovel will often get stuck in it when you're taking out the bedding. Concrete floors hold moisture, and during a humid summer, they can make an already damp goose area swampy.

I enjoy keeping geese in a stall finished with floorboards. Inexpensive, rough-sawn hemlock can be laid down, or may already be there if you're converting an old barn. The issues with floorboards are numerous: they will eventually rot, especially with water-loving geese living on them, and they can also be eaten away by insect infestations. Sprinkling diatomaceous earth will help with the second problem, and regular cleaning delays the onset of the first. However, if floorboards are an option for you, I would recommend them, because cleaning bedding off boards is easy and smooth, and they won't hold in moisture or cold like concrete does.

Fencing

Larger farms often allow their geese to free range, allowing them to forage for food, and if you are using them for guarding purposes, this plan can be helpful. But on a small farm or in an urban setting, free ranging is impractical. Fortunately geese tend to be easy to fence. Given enough space, they will be quite comfortable within a paddock. In a too-small enclosure, they

Fenced pasture space with free access to food and water.

can eat all of the vegetation fairly quickly, but supplementing with free access to grain and regular feedings of greens can keep them quite happy.

An outdoor pasture should allow at least ten square feet per bird. Most geese are not particularly avid fliers, so three- or four-foot fencing will keep them enclosed. If you have trouble with flighty geese (some lightweight breeds like Romans may be more apt to fly) or are concerned about airborne predators, taller fencing or a net over your paddock may be needed.

Geese can be easily contained with four-by-four-inch square wire fencing, also known as stock fencing. Electric fencing is not necessary to contain your geese, but it is recommended to keep predators at bay. If you are not concerned about predators and can afford the cost, wooden fencing also keeps geese in their pasture effectively. No gaps in your wire or wood fence should be larger than six inches, or a goose may sneak through.

Water

Geese love to swim, but they do not need a pond to be happy. As waterbirds, they love the opportunity to thoroughly bathe. They can maintain their feathers without much water, but preening in water helps keep their feathers shining and waterproof. A rubber tub is often sufficient for them to wade and bathe in and is especially reasonable in freezing winter weather, being demanding for the goose keeper to keep free of ice.

A kiddie pool works well for geese during summer, often providing enough water for them to mate.

Geese can mate without water, but they prefer to do the deed while swimming. If you plan to breed geese as part of your farm, having a pond will improve your fertility rates and number of matings. They do a complete courtship dance on the water that they can mimic in a smaller pool but often are fussy and may abstain.

If you do have a pond, your geese will go to it daily. They will love being able to groom and dive and nibble at small grasses and algae that grow along the water's edge. Simply put, geese do not need water, but they do love it.

For grooming and swimming, geese are satisfied with a small substitute for a pond. When it comes to mealtime, they all need plenty of fresh clean

Geese love to swim from an early age.

water for drinking. They must be able to submerge their bills in order to keep their air passages clear, and clean water is important for them to stay healthy. A large plastic chicken waterer provides them with enough room to fully submerge their beaks, and it can be scrubbed out easily when the water becomes dirty.

During the winter, a heated dog water bowl will keep your geese happy, as long as it is regularly cleaned and refreshed. Providing geese with fresh clean water regularly is one of the most important steps towards a maintaining a healthy flock.

Feed Requirements

Goslings and young geese require free access to feed or crumble to give their growing bodies the necessary proteins. Geese can eat wheat, pellet feed, or crumble feed as their primary food source. Goslings up to two weeks old require about 23% protein in their feed; up to 6 to 8 months, they need 19% protein; and once they are laying regularly, need 17% protein. They also require many other minerals and vitamins that are found in specially formulated bird foods.

Chick starter feed is not suitable for goslings. Unlike chickens, geese need a certain amount of the niacin, a B vitamin, which they usually

A free range goose flock.

Goslings up to 4 weeks	Goslings 4 weeks to 6 months	Geese 6 months and older
Protein: 22.5% kcal/kg	Protein: 18% kcal/kg	Protein: 15.5% kcal/kg
Lysine: 1.16% kcal/kg	Lysine: .90% kcal/kg	Lysine: .80% kcal/kg
Calcium: 1.00% kcal/kg	Calcium: .90% kcal/kg	Calcium: 2.50% kcal/kg
Phosphorous: .40% kcal/kg	Phosphorous: .35% kcal/kg	Phosphorous: .30% kcal/kg
Vitamin A: 15 KIU/kg	Vitamin A: 15 KIU/kg	Vitamin A: 12 KIU/kg
Vitamin D: 3 KIU/kg	Vitamin D: 2 KIU/kg	Vitamin D: 2.5 KIU/kg
Niacin: 55 mg/kh	Niacin: 55 mg/kg	Niacin: 40 mg/kg
Potassium: .83% kcal/kg	Potassium: .80 % kcal/kg	Potassium: .80% kcal/kg

Even geese on pasture love special treats.

cannot get from standard brand chicken feeds. A lack of it leads to leg problems and stunted growth in goslings and adult geese, as well as an ailment known as angel wing. Because ducklings also require niacin, many feed stores carry crumbles formulated specifically for ducklings and goslings that contain it. Be sure to check before using it as your primary feed. You may also supplement niacin by adding a tablespoon of brewer's yeast sprinkled over their feed every day.

There is another reason that chick starter should not be fed to goslings. Chick feed contains medication to combat a condition called coccidiosis, which can affect waterfowl occasionally, but more commonly this medication may harm goslings. Studies on these medications have been inconclusive, and many farmers do feed medicated chick starter to their goslings, but it is recommended to provide non-medicated feed as a precaution. Non-medicated chick feed that either includes niacin, or has niacin added to it by using brewer's yeast, is the allowable feed to offer goslings. Adult geese, similarly, should have their protein needs met either by chicken feed supplemented with brewer's yeast or special waterfowl feed that includes niacin.

Feeding Costs

One of the primary reasons to consider geese for your farmyard is that, for the number of potential benefits, they are remarkably inexpensive to keep. The amount of grain geese consume varies depending on how much pasture they have, but with even minimal supplemental feed, they consume relatively limited amounts of grain.

A pair of adult geese with no pasture or supplemental food can eat a half a pound of feed a day. That would allow a 50-pound bag of goose feed to last 100 days. Geese kept on pasture will eat a quarter of that

In wintertime, geese will eat more grain because they don't have access to pasture.

or less, and with plentiful greens, they need not be offered any grain at all. The overall cost of feeding a goose should not be more than a few dollars a day, which is easily recouped if you sell its eggs or meat, both delicacies that can sell at a premium.

Young geese consume less feed than adults, but they do need dedicated grower crumble designed for goslings. These feeds can be slightly more expensive; however, you should only have to buy a single 50-pound bag for a pair of goslings that will last them until they are old enough for adult bird food.

Larger flocks, of course, consume more; a flock of ten goes through a 50-pound bag within a week. However, the potential profits from your goose farm increase with more geese, whether you are using them for weeding, eggs, or meat.

Fields and Pastures

As dogs love bones and ducks water, geese love grass. Fairly strict vegetarians, they will occasionally eat bugs and mealworms, but their favorite foods are fresh greens; whether lawn grass, weeds, lettuce, or your garden vegetables, greens are the yummy first choice for all geese. This is why many farmers use geese for weeding, especially around crops that they will not nibble on, such as in orchards and vineyards.

If your geese free range and you keep a garden, you might want to fence off the garden area. It may take a while for geese to discover your lovingly cultivated home-grown greens, but when they do find them, it takes only a few minutes for their long bills and ravenous appetites to decimate your crop. They may be very happy, but you will not be. Most geese can be fairly particular; they have favorite vegetables and others that they turn their

noses up at. Similarly, I've known some geese who love mealworms, and others that will not even try them.

With a large lawn or yard to roam in, they'll keep themselves thoroughly amused nibbling on grass and dandelions. If they are contained in a smaller paddock, it is a good idea to offer your birds these regularly as a treat. Heads of lettuce, lawn clippings, and the greens of root vegetables are excellent fare to keep them happy, content, and provided with the necessary nutrition.

There are some foods you should never feed your geese. Nightshade plants, from fruit to greens, are poisonous for all poultry. These include potatoes, tomatoes, and eggplant. Any food that has gone bad or is moldy is not good as feed, and junk food is also bad for geese (including chocolate).

Plants Poisonous to Geese

Along with potatoes, tomatoes, and eggplant, some other flowers and vegetables are toxic to geese. Some of the most common examples are foxglove, rhubarb leaves, honeysuckle, ivy, and rhododendron. One of the most toxic, ironically, is the gooseberry, which contains high levels of cyanide.

Because they are vegetarians, stick to vegetables, fruits, and greens. Our geese love watermelon, and some are known to enjoy onions and garlic; however, their eggs may take on a nutty flavor if they have too much of either.

Chapter Four:

Health and Illnesses

W E DO OUR BEST TO KEEP OUR GEESE HEALTHY at all times, but sometimes illness strikes and it is important to know how to treat your geese when they are sick. The best steps towards a healthy flock are preventative, making sure your birds have a healthy diet and safe, sanitary living conditions.

Some of the most common illnesses that geese can suffer are detailed below, along with information on how to treat them. The best thing to do if you have a sick goose is to take it to the vet for a proper diagnosis. When you

A flock of healthy geese have bright eyes and active characters.

first get your geese, take the time to find a good avian veterinarian in your area. Having that connection can make all the difference for your birds.

Understanding the birds that you are keeping, their breed and normal habits, will help you to recognize anything unusual. You want to be able to notice a malady as soon as an issue begins, and quick treatment of diseases is the best hope at helping your birds survive them.

Angel Wing

Symptoms

Angel wing is a relatively common ailment affecting domesticated waterfowl. Also called "drooped" and "dropped" wing, it causes the flight feathers to grow out at twisted angles from the wing joint, often sticking straight out from the sides of the bird. Angel wing inexplicably occurs much more frequently in the left wing, and is more common in males than females. The odd angles of the flight feathers will render a goose totally flightless. Angel wing develops when a bird is young, before sixteen weeks of age.

Causes

Angel wing is most frequently caused by dietary factors. In geese, it almost never occurs in the wild, except where they are fed regularly by people instead of foraging on their own. A study by Janet Kear, Senior Scientific Officer at the Wildlife Trust in the UK, concluded that angel wing is most likely caused by an excess of high-protein feed, which make wings grow unusually fast, thus becoming overly heavy. Other studies have indicated that a lack of certain vitamins, including vitamin E and calcium, contribute to angel wing disease.

Treatment

Once a goose displays angel wing, it is too late for any treatment, as this disease is incurable. Fortunately, angel wing causes its sufferers no pain or ill effects other than the inability to fly and an unsightly lopsided appearance. This may not be a critical factor unless you plan on showing your geese. The best treatment is, of course, preventative. Care should be taken to give young geese the right amount of protein and vitamins in their feed, and not to try to hurry their growth with high-protein feed. After three weeks of age, goslings should be fed no more than 18% protein and should be given feed that contains about 1% calcium and 20 milligrams of vitamin E. Any commercial product, purchased at your local livestock feed store, will have a label illustrating the nutrition enclosed.

Aspergillosis

Symptoms

Aspergillosis is a fungal disease that can affect all types of poultry. Because this fungus attacks the lungs, it is most commonly noticed when birds have difficulty breathing, and take gasping, labored breaths. Birds with aspergillosis have rattling respiration and become very lethargic. While aspergillosis can occur in any bird, it most commonly affects goslings when they are very young.

Causes

As a fungal disease, aspergillosis is caused by an unclean brooder or coop.

Treatment

To prevent aspergillosis, regularly clean your birds' living area. If you are hatching goslings, make sure that the incubator is clean before beginning the process. The eggs you are using should also be clean. Moreover, make sure you have a fresh, clean brooder ready to put your goslings in. If you notice signs of aspergillosis, remove all bedding and disinfect the living area using 1:200 copper sulphate (source: FAO Animal Production & Health Paper, Buckland & Guy, 2002, ISSN 0254-6019). Birds suffering can be treated with antifungal remedies such as amphotericin B; however, treatment is not always effective once they are displaying symptoms.

Bumblefoot

Symptoms

Bumblefoot is an infection on a bird's foot that swells and becomes a bulbous protuberance that prevents it from walking comfortably. You will most likely notice bumblefoot because one of your geese is lame, and upon inspection, the pad of the foot will have a round lesion or other swelling.

Causes

Bumblefoot is caused when a bird has an open wound on their foot, often just a small scratch that might be caused by a pebble or something else in their run. The bacteria that causes bumblefoot (*Staphylococcus*) infects these small wounds, resulting in swelling and discomfort.

Treatment

Bumblefoot can be treated with veterinary prescribed antibiotics, but it is often necessary to lance the bulbous infection. Using a sharp scalpel, slice the round area and drain the puss carefully, then bind the wound with padded gauze to prevent infection. Geese being treated for bumblefoot should not be allowed to swim, and their dressing will need frequent changing. You may also soak the drained wound in Epsom salts before covering. Keeping a clean and tidy area for your birds will help to prevent bumblefoot infection, as will treating the wound with an antiseptic wash before it becomes infected.

Coccidiosis

Symptoms

Coccodiosis is most common in goslings between 3 and 12 weeks old. There are two different strains of the disease, one renal and one intestinal. An infected gosling will be lethargic, unsteady on its legs, have diarrhea or white feces, and refuse to eat.

Causes

A parasitic infection, coccodiosis in geese is caused by either *Eimeria truncata* or *Eimeria anseris*, and is most commonly the result of poor hygiene or lack of ventilation in the goslings' brooder.

Treatment

Coccodiosis does have a very high mortality rate; however, it can be treated with sulphonamides. Many feeds designed for baby chicks contain a coccidiostat to prevent this disease, and these can be offered to your goslings as long as they meet their other nutritional requirements. Adding a tablespoon of apple cider vinegar to a gallon of your goslings' drinking water will help prevent infection, as will maintaining a clean, dry area for them.

Derzy's Disease

Symptoms

Highly contagious, Derzy's disease is a virus that affects geese and Muscovy ducks. It occurs in ducklings and goslings under 5 weeks old, and can be noticed with lack of interest in feed, excessive water intake, profuse diarrhea, and discharge from their nostrils.

Causes

Derzy's disease has also been called goose plague, goose hepatitis, and goose influenza. It is caused by a parvovirus and transmitted through infected water or feed, or by contact with the feces of infected birds. Adult birds can be latent carriers, but there are vaccines for this disease available to prevent adults from infecting goslings.

Treatment

There is no treatment for Derzy's disease, and it typically has a 100% fatality rate in very young goslings. Older goslings will often survive the disease.

Egg-binding

Symptoms

Egg-binding can occur in females of any egg-laying species, including geese. Egg-bound geese will show signs of depression and lethargy, and will often sit in the nesting area straining to pass the egg for some time.

Causes

Egg-binding is caused when an egg cannot, for various reasons, pass through the oviduct to be expelled. It is most commonly caused by a poor diet, one with insufficient calcium and vitamin D. It also is more common in birds laying their first eggs.

Treatment

To treat an egg-bound goose, offer her a bath of warm water to relax her muscles and massage the area to help her pass the egg. Applying K-Y Jelly to the cloaca will help. To prevent this issue from occurring, feed your geese a nutritious, calcium-rich diet.

Mycoplasma

Symptoms

A bacterial infection, mycoplasma is a common disease for many birds, including geese. It is highly infectious, so one bird can quickly transmit it to an entire flock. Infected birds will have discharge from their nostrils and eyes, act lethargic and slow moving, and can display some lameness. The most common sign of mycoplasma, however, is a lack of egg production.

Causes

Mycoplasma affects the respiratory system, and it is usually a slow-onset disease, which can make it hard to diagnose. Geese can catch it from infected chickens or other poultry, or from dirty living conditions. The disease can also be passed down from parent to gosling.

Treatment

Mycoplasma is chronic but not often fatal, and symptoms can be treated with tilmicosin or tylosin in the bird's drinking water. Ensuring that your breeding stock is free of the disease and maintaining a clean coop for your geese and all your other fowl will help to keep your farm safe from this disease.

Paratyphoid

Symptoms

Paratyphoid, also called salmonellosis, is an important disease to be aware of as salmonella infection can be transmitted to humans. Typically, paratyphoid will affect goslings under 6 weeks old, but it also can infect eggs in the nest, making it crucial that you collect your goose eggs regularly and wipe them clean before consuming them. Infected goslings will be lethargic, drop their wings and heads, and have messy diarrhea.

Top left: *Author holding two goslings.*

Bottom left: *Spending plenty of time with young goslings will ensure they imprint.*

Top right: *This young African gosling has plenty of attitude.*

Bottom right: *Gosling on pasture.*

Top: *Gosling interacting with grown geese.*

Bottom right: *A four week old Dewlap Toulouse goose. Personalities start to become clear in geese at an early age.*

Raised together, goslings will bond with each other and be inseparable throughout their lives.

Center left: *Geese enjoy bathing in snow, but they are not fond of cold weather.*

Center right: *The distinct blue eyes of Rupert the Sebastopol gander.*

Bottom left: *Geese form close bonds with each other. This pair, a Dewlap Toulouse and a Buff, have been together since hatching and remain best of friends, even though they are two ganders.*

Left: *Our first geese, a mated pair of African geese.*

Right: *Geese form bonds with humans also.*

Rupert the Sebastpol is a beautiful sight, cleaning his feathers in the summer sunshine.

Top: *A gander like Pete will take 2–3 geese as mates. It is important to keep a healthy goose to gander ratio to avoid fighting between your geese.*

Bottom left: *Pete the Roman Tufted gander is one of our most closely bonded geese.*

Bottom right: *While most geese cannot fly, they still enjoy stretching their wings and showing off.*

Top: *Mother goose with goslings.*

Bottom left: *Some ganders will bond with each other — like Precious and Arnold.*

Bottom right: *Geese with goslings — geese co-parent unlike most birds.*

I can watch geese fastidiously clean and arrange their feather all day.

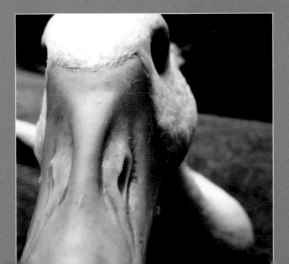

Center left: *One of the best reasons to keep geese is for their delicious, giant eggs.*

Center right:*Chicken egg next to goose egg for size comparison.*

Bottom left: *Geese are very curious creatures, and Brutus is particularly camera-friendly.*

Precious is the epitome of the attitude that geese are famous for.

Center left: *Geese with chickens — guarding, co-existing.*

Center right: *Geese in pasture — geese will enjoy as much space as you are able to offer them.*

Bottom right: *Arnold our Buff gander at only a few weeks of age.*

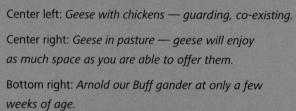

Causes

Like many infections, paratyphoid thrives in unclean conditions. A clean brooder and nesting area will go a long way in preventing this disease.

Treatment

Paratyphoid in goslings can be treated with antibiotics. Maintaining sanitary conditions will prevent an outbreak from occurring.

Spraddle Leg

Symptoms

Spraddle leg affects goslings usually only a day or two old. It is most often found in hatchery stock or birds hatched in an incubator on a smooth, slippery surface. It also occurs in goslings who are not given feed containing enough niacin. Birds will be unable to hold their feet together, and if they are able to stand at all, they will have a squat, unstable position.

Causes

Apart from a smooth hatching surface, the other cause for spraddle leg is niacin deficiency. Niacin (vitamin B3) is critical to the healthy development of goslings.

Treatment

Provide a rough surface for your goslings to stand on and consider splinting affected goslings' legs with veterinary wrap. To prevent spraddle leg, and to treat affected goslings, make sure that they are receiving at least 55 milligrams per kilogram of niacin in their diet. Niacin can be supplemented by adding brewer's yeast to goslings' feed.

Tapeworms and Roundworms

Symptoms

Tapeworms and roundworms (or nematodes) can affect any poultry, and often you will never notice their presence. Infected geese will be lethargic and act depressed. There are types of worms, however, that can be detrimental to your goose's health. It is almost impossible to fully diagnose worms in geese without an autopsy.

Causes

Tapeworms and roundworms are naturally occurring, and it is possible that your fowl have them and show no symptoms at all. Certain types of these worms can have negative effects, and they are usually first contracted by geese drinking pond or stream water.

Treatment

If you suspect that your goose is affected by worms, a few drugs have been experimented with for treatment but there is no single cure. The best way to avoid them is by rotating pastures and keeping a clean and sanitary barnyard.

Other Health Issues

Geese can be affected by a whole host of possible health issues, many of which are curable with responsible management. They can become stressed, especially in new living conditions or if they are being harassed by a dog or other predatory threat. Stressed geese will not lay, have messy feces, and often won't be interested in food. Remove the stressor and your flock will return to full health.

Geese can also suffer exhaustion and heatstroke. This is easily preventable by providing plenty of fresh, clean water and shade on hot days. If you notice your geese holding their beaks open to breathe and carrying their wings away from their bodies, they are likely too hot and would appreciate some cool water.

African and Chinese geese can get frostbite on their knobs during cold winter weather. Providing a dry, wind-free area for them prevents frostbite, but if it does happen, you can treat the affected area with an application of petroleum jelly. It will display as spots of pink or yellow on the otherwise black or orange knob.

As you may have noticed, the majority of diseases that affect geese are caused by unclean living conditions. Regularly cleaning your coop and providing fresh, dry bedding is crucial to your flock's health, and it is especially important that newly hatched goslings are raised in sanitary conditions. Poor diet is another common cause for illness in geese.

Allowing your birds their natural inclination to forage keeps them healthy and thriving. Make sure you know that the feed you offer them has the adequate amounts of necessary vitamins and minerals to maintain good health. Read the label. While the list of possible diseases your geese might contract may seem long and daunting, following these two simple steps, providing good nutrition and maintaining a clean home, can keep your flock happy, healthy, and disease-free.

Chapter Five:

Primary Predators

*Y*ES, IT'S TRUE THAT GEESE MAKE EXCELLENT GUARDIAN ANIMALS, but there are also times when they can be preyed upon, and as a keeper of geese, it is your duty to keep your birds safe. There are many predators that may threaten your flock, so basic precautions such as providing strong fencing and secure nighttime quarters go a long way in preserving their welfare. A large flock is less likely to be attacked then a single goose or a pair; a group of birds are significantly more intimidating. There's safety in numbers. That being said, you can take certain steps to keep your geese, whether a single companion or a gaggle, safe from predators.

Coyotes, Foxes, and Big Cats

The most common potential predators of geese where we live in Maine, and throughout most of North America, are large carnivorous mammals, such as coyotes, foxes, and big cats. While foxes and some of the wild cats may be too small to carry off a fully grown goose, they'll still help themselves to goslings and young adults. Coyotes can be surprisingly fearless and are ready and eager for any opportunity to carry away a full-sized goose from a farmyard.

Fortunately these animals, being nocturnal, prefer to hunt at night. You might see foxes and coyotes during the day, but unless the target is too easy to pass up, they usually bide their time until nightfall. This is one of the reasons that staying active on your farm is helpful: the more that predators like this see people around, working and protecting the animals, the more they will think twice about attacking.

It can be easy to tell if you've got a coyote pack nearby. Their yipping cries are bone-chilling, and as I have experienced, they are frequently heard at dusk, often seeming only a few short yards from your home.

The best way to protect against large mammal predators is with a solid fence and a secure night area. Since these predators prefer nocturnal hunting, if your birds are sheltered and secure at night, they are largely out of danger. With an active homestead, they may be safe wandering from farmyard to dooryard during the daytime, especially if you have a large flock, but if you are not around or keep only one or two geese, keep them within a secure fence.

A good fence should be at least four feet high, with no-climb fence braiding (known as "stock fencing"). A strand of electric wire, if the rest of the fence is non-electric, further helps to keep predators at bay. Make sure that your fence goes all the way to the ground. On top of that, consider burying the first few inches to discourage the enemy from digging under and getting inside.

At night, the only truly safe place for a preyed-upon animal is inside a building, be it a coop or a barn. Four walls and a roof, with a door that latches and a solid floor, make it difficult for a four-footed predator to attack. If you are not able to provide this kind of shelter, at least ensure that your birds are enclosed on all four sides and that the entry door closes securely.

Other ways to keep predators away at night include playing a radio or positioning flashing lights, and, of course, the tedious task of regularly checking on your geese. Both the radio and lights will often keep threats away, but some predators will figure out that these are red herrings and not truly dangerous, and so without fear, they very well may return to harass your flock another night.

Birds of Prey

There was a time when I didn't consider that birds of prey would be an issue with geese because geese themselves are so large, then we lost a beloved and fully mature goose to a bald eagle. Our adult male African goose was close to 25 pounds, but when the determined eagle carried him away, there was not a trace remaining.

Most of the smaller hawks and falcons are no match for a goose. One of the predators most difficult to combat, the American bald eagle swoops down from the sky, and no amount of fencing along the ground perimeter is going to stop an attack. Geese have excellent vision and always have one eye to the sky, which means they can be great alarms should smaller raptors, such as hawks, threaten your chickens. But even a gallant goose cannot evade the stealthy wings of a mighty predatory bird like the eagle.

Often the best way to keep eagles at bay is to keep your flock relatively close to home. Fencing can help keep a flock near the barn and other buildings, which can deter eagle attacks. Keeping geese and other farmyard fowl sheltered at night will protect them from attack by nocturnal owls, another bird that can grow large enough to carry off a goose.

Unfortunately without a covered pasture or investing in a protection animal, like a livestock guardian dog, preventing eagle attacks can be difficult. Fortunately there are many farming areas where birds of prey are limited, and often various raptors are kept well-fed by fishing nearby waterways.

Vehicles

Vehicles are not truly predators, of course, but cars and trucks can be a major threat to geese who have no idea that they are capable of inflicting

Geese seem to love open roads.

injury. Geese will try to stand down a car, a battle they will always lose. My geese habitually favor having a morning nap in the middle of the road. We moved our flock to new homes, but still they find their way to the road around midmorning, preen themselves, and then settle down for some quick shut-eye. No amount of chasing averts them.

The only real protection against traffic is to keep your geese away from the road. Good fencing is the way to go, and even if you otherwise allow your flock free range of your farm, a fence to prevent road access is an excellent strategy. If you have many acres, it is possible to simply house your geese away from traffic, but this is a luxury not all of us have.

Weasels, Skunks, Raccoons, and More

Small predators are opportunists, and they certainly will take on a big target like a goose if it is injured or otherwise looks like an easy meal. But the main concern with skunks, weasels, and raccoons is the harm they can bring to goslings and eggs. Raccoons, skunks, and even possums will bite a small incision in the egg and suck the contents out, leaving just an empty shell. Harm can come to a mother goose defending her eggs or hatchlings from such an onslaught, and even if the goose escapes danger, losing your egg harvest is a disheartening setback.

It is hard to protect against smaller mammal predators. Weasels are especially persistent, and able to sneak through small holes that you might not even notice. Once again, proper night shelter is the best defense. A solid building will fend off most pests. Search out meager holes you otherwise might have overlooked, and when you find one, plug it up before vermin become a nuisance. Use plywood or, if they're quite small holes, hard wire mesh. Like larger predators, skunks and raccoons prefer hunting at night, so if you provide a tight night shelter for your geese, they can sleep in safe quarters, guarded from the damage and trauma of attacks.

Unlike coyotes, these predators are solitary creatures with small families, if any at all, and they do not travel in large packs. Because of this, a diligent farmer can deter small predators by daily scrutinizing of the surroundings and taking proper precautions against trouble. If you are challenged by the unwanted attentions of a weasel, set a trap, relocate the animal, or destroy it. This can take time, but eventually, you can remove the problem entirely.

An alert goose, especially if she is protecting her nest, can seriously harm a small predator all by herself. Geese will help protect chicken flocks from skunks and weasels large enough to carry off a young hen. The exception to this is during the darkness of night, when geese can barely see and often won't notice trouble until it is too late.

Dogs

Beware a loose dog, for one can decimate a flock of geese. Unlike most wild predators, dogs will often kill their prey indiscriminately and leave the carcasses behind, instead of eating them or carrying them off. Even a large goose is no match for a full-size dog.

Livestock guardian dog.

Dogs on the homestead should be carefully and attentively introduced to any new livestock animal. It is wise not to allow dogs off the leash around your geese, especially if the dog is unpredictable or is a high-energy animal. Strict, attentive correction is due if dogs attempt to chase geese. This necessary training allows the dog to adjust to the new birds. A well-trained dog can become fond of poultry on the farm and can learn to help to protect them. Livestock guardian dogs, breeds that have been raised for generations to protect livestock, can be hugely beneficial.

The greatest canine risk to your flock of geese may be neighborhood dogs or strangers in the area. While your own dogs may be well-behaved and carefully monitored, you never know when someone down the road might let their dog loose without realizing that it will head right for your farm. We have lost plenty of chickens this way, thanks to irresponsible neighbors. Once again, the best defense is good fencing and vigilance. If you observe a neighbor's dog loose around your property, let the owner know that you are concerned for your birds. A solid fence will stop a dog, and unlike other predators, it is rare to have a neighborhood dog attack during the night.

The primary predator of geese is, of course, people. Plenty of us raise geese for their meat or even hunt them in the wild. If you are raising geese for food, you want to keep in mind that you are the predator who will reap the harvest. If you're keeping geese as pets, or for their many other benefits, make a commitment to maintaining a good fence and a secure shelter. Such precautions will ensure that the birds remain unharmed and thrive as part of your family for many years.

Chapter Six:

Keeping the Neighbors Happy

*K*EEPING GEESE IN AN URBAN ENVIRONMENT can be a challenge. They are large birds, with wingspans that measure five feet or more, and as such they need space to stretch those wings. The proper pasture with plenty of room, amusements, and food will help to keep your geese happy even when your neighbors are close by.

Geese love to wander if they are not kept in a pasture space.

Fortunately most breeds of geese are incapable of flying to the neighbor's or further afield. The majority of domestic geese have bodies heavy enough that even their vast wingspans cannot get them more than one or two feet off the ground. Even totally flightless birds do like to make a show of running, wings outspread, when they find a wide open space. If you keep your geese in a run, the most you should see of this behavior is a daily display of wingspan by beating them and, frequently, calling a celebration of his or her beauty.

Space Requirements

Whether you're in town or the countryside, outdoor space requirements for geese is an important consideration. Adult geese require pasture that allows about 10 square feet per bird. In a backyard urban area, you might

Geese need room to spread their wings.

be tempted to scrimp on giving up so much real estate to your birds, and while 10 feet per goose might seem to overreach what you have to offer, keep in mind that providing adequate space means that your geese will not only be more content, but they will stand a better chance of maintaining good health. Crowded animals are more prone to being aggressive to the other animals they find themselves confined with. The tighter their space, the higher the possibility of injury inflicted on one of the birds, especially one at the bottom of the pecking order. Birds injured in such situations suffer from wounds that might easily become life-threatening.

Another small-space issue with geese is that when kept in compact quarters, they will quickly eat their vegetation down to the bare ground; however, that isn't as much of a problem if you keep them happy and amused with a dependable supply of treats. On the in-town backyard farm, the best way to keep geese is often a combination of a smaller fenced area with a gate opening into a yard where they can wander when supervised. Geese do have a tendency to wander, so when you can't be there to keep at least one eye on them, then it is probably best to hold them in their primary pasture, not only for their own protection, but in case they get curious about what's next door, or across the street. Besides the dangers of cars in the road, the neighbors might have a dog that would enjoy chasing and catching a stray bird. Don't let it be yours.

Wandering and Exercise

Never forget that geese are naturally curious creatures. It is possible that a flock of geese will stay close to home, hanging out right around their own yard all day and show no interest in walking off. But if they hear some activity a couple of houses away, or the sound of neighbors talking across the road, they will more than likely want to investigate. This is part of what

One Goose or Two?

Getting just one goose isn't recommended. Geese are flock animals, and they need someone or something to bond with. There are legends of single geese becoming overly attached to wheelbarrows and watering cans in the absence of another animal. One goose can be a great companion if you have plenty of time to spend with them in place of a "mate," or if you have a flock of ducks, chickens, or even goats or a horse that they can bond and talk with in place of another goose. However, if you're looking to limit the number of geese on your farm, consider getting at least two so that they can have each other.

makes them good watchdogs, but it can also be a challenge in an urban setting, and that is why I recommend keeping your geese in a fenced pasture. If and when you are concerned about them not getting the chance to stretch their legs enough, then letting them wander on the yard when you are home and able to watch them is a good option. Chances are, they will prefer staying close to you and your activities and often are like shadows following just behind you all day.

Entertainment in the Run

If you cannot let your geese out occasionally to wander, you can make their confined spaces more interesting for them. In confinement, geese will quickly eat down the local vegetation, and a pool for swimming can only keep them amused some of the time. You can build toys for them to play with that might help to stop them gnawing on the fence or coop in

With enough space, geese and chickens can happily mingle.

boredom. A handful of treats, loose lettuce, or lawn greens will delight them for a few hours. A flake of feed hay regularly tossed inside their fence may not get completely consumed, but they will be fully entertained simply from chewing around the dry stems. Geese also enjoy conventional dog toys or bells to chew on and play with. Nosing around a few toys will not only help keep them amused, it will also keep them relatively quiet. You can also hang a head of lettuce from a rope or in a wire basket inside their run. This not only provides a tasty and nutritious treat, but keeps them focused on the swinging action, and busy figuring out how to get the food out.

Noise and Neighbors

Another thing to consider regarding geese on the urban farm is their noise levels. Geese will honk in various situations. For example, if they see anything out of the ordinary, or if they are ready for a meal, or when it is time to come in or out, you'll hear them telling you (and everyone else) about it. They also sound off as part of normal social communication through the flock. Many geese have high-pitched honks that can be heard for some distance. Although their voices are by no means a reliable method of determining the sex of your geese, the male geese often have a more ear-piercing honk than females, whose voices are usually deeper and less shrill.

To limit the noise your geese make, you can select breeds that are less vocal, keep a smaller flock, or simply try to limit yourself to females. Adding toys to their enclosure can also help to keep them quiet. To promote both their mental health and good behavior, decide to spend plenty of time with them, if for no other reason than your companionship may help to limit their vocality. Geese are smart animals, and they love being able to hang out with "their people."

The Mixed Flock

Having other birds with your geese will keep them mentally stimulated, but it isn't always the best way to keep them quiet. Since geese

Town Ordinances

It is crucial to look into your town's rules and regulations before adding any livestock to your home. Doing your research can prevent fines and the difficult and heartbreaking act of having to get rid of animals that you've already bonded with. Visit or call your local town office to see if they have any restrictions on geese or poultry.

are naturally protective, they'll be more likely to honk at any suspicious signs of trouble for their duck or chicken friends.

If you're planning to keep other poultry mixed in with your flock, then having only a few geese is a good idea. A large number will bond together and might bully the smaller birds, whereas one or two will co-mingle with the rest of the group. Geese kept in a pasture with chickens or ducks are usually harmonious companions. They are not as aggressive with their pecking orders as chickens are, although they like to establish a hierarchy and then reinforce it occasionally with gentle, firm nips.

Sometimes geese will be more aggressive towards ducks or chickens, and I like to provide my other birds with areas where the smaller ones can get away from the geese if necessary. I have never feared for my birds'

safety, but it is helpful to provide ducks and chickens with a defensive area where they can relax. Since geese are such big birds and they do not roost as chickens do, establishing a "getaway" space is usually not difficult. An area inside your coop that is walled off with a doorway too small for a goose to pass through, but large enough for chickens and ducks, can provide a safe space, and you can use a similar method in your outdoor run. Roosts are also handy devices for chickens to get out of the way of geese if they want to.

For a farm with plenty of land and pasture space, keeping geese can be remarkably undemanding. It is still worth your time, even in the wide open spaces of a country farm, to spend time with them your birds, especially when they are goslings. Your attention and companionship will ensure that they will not be aggressive, as they remain comfortable and familiar with and around people. In terms of harassing a next-door neighbor, there is less to worry about on a large expanse of farmland.

On a small farm, or in an urban or suburban setting, you'll have to do more to make sure your geese do not bother the neighbors. I would recommend keeping a minimal flock, no more than two or three geese, on an urban farm. There are two reasons for this. One is that a smaller flock will be happier in a modest, compact space, while a dozen geese on a one- or two-acre homestead are likely to be crowded and discontent. The other is that two or three geese are quieter, and they're more likely to get along with your other poultry.

When I kept a pair of geese with a flock of chickens on a one-acre farm in a suburb of Portland, Maine, they caused almost no problems. Occasionally, they would start honking all at once, and I felt anxious that the noise would disturb the neighbors. But loud outbursts like that were rare, and we never had complaints about the noise. Notably, as soon as our flock expanded to four and more geese, every day presented problems. Left to free range, my flock would invariably end up in the road or foraging in the flower pots on our neighbor's porch. I tried keeping them penned in, but being locked in their run full time caused them to complain noisily all day. In addition, the run had originally been built for just the small clutch of chickens, not really big enough for both the geese and the hens. There

were more geese than hens, so within the cramped space, the geese continually bullied the hens. Yes. There was trouble and frustration for both the birds and the keeper.

Besides the danger of wandering into the road, our geese left calling cards (goose droppings) on the neighbor's back porch. In contrast, if I was outside working in the garden, the geese stayed right at my side the whole afternoon, never venturing far afield and rarely making too much noise. If they had been raised more accustomed to the confines of a fenced-in pasture space in the company of chickens, and if I had not expanded the flock so quickly, they would have been more comfortable and more manageable had I provided regular, supervised outings in the yard.

Nowadays, on a larger farm, my geese are very much homebodies. This is partially because they have fewer stimuli. There are no neighbors talking next door, joggers moving past the house, or dogs barking in the distance. There's not much activity in the surrounding fields to attract their attention. Reducing distractions is certainly one way to keep a flock docile. Sometimes I increase the entertainment value by offering them a flake of hay in their shelter or a pile of fresh greens, to maintain a peaceful, content flock.

Chapter Seven:

The Farmer's Alarm System

$\mathscr{G}$EESE HAVE BEEN USED AS GUARDIAN ANIMALS FOR CENTURIES. Why? Because they are uniquely suited to the task. Not only are their piercing honks excellent alarms, but they are instinctively protective and more naturally suited to guard work than humans, or even dogs.

Because birds can see ultraviolet light, their vision is far superior to that of people. Their distance vision is remarkably good (the better to spot possible predators), and they can sense movement long before an

Geese have a natural talent for guarding your farm.

ordinary person can. While most birds don't react to what they see in a way that is helpful to people, geese do. Geese sound off loudly and aggressively, ideal attributes for guarding. Geese are also territorial: they know where their home is, and they defend it, especially during mating and hatching season. Few domesticated birds share the same territorial tendencies, and fewer still have the pugnacious attitude for attacking intruders.

The History of the Guard Goose

The first example of geese being used for protection dates all the way back to Ancient Rome. Geese were sacred to the Goddess Juno, and so the flocks that lingered around her temple were left untouched by locals that might otherwise have eaten them. In 390 BC, Rome was under siege by the Gauls. Late one night, the Gallic army attempted a surprise attack by scaling the Capitoline Hill, where the temple of Juno was located. The local dogs did nothing to stop them, as they were easily bribed with fresh cuts of meat. But the flock of Roman Geese — a breed still surviving today — noticed the disturbance immediately and began to call loudly. Here is some of Plutarch's description of the events, from his work *Parallel Lives:*

> About midnight a large band of them [the Gauls] scaled the cliff and made their way upward in silence .... Neither man nor dog were aware of their approach. But there were some sacred geese near the temple of Juno .... The creature is naturally sharp of hearing and afraid of every noise, and these, being specially wakeful and restless by reason of their hunger, perceived the approach of the Gauls, dashed at them with loud cries, and so waked all the garrison .... The defenders, snatching up in haste whatever weapon came to hand .... So the Romans escaped out of their peril.

This heroism of the geese did lead to a rather disturbing annual festival that included a display of geese dressed in purple and gold, but focused primarily on shaming the dogs for not sounding the alarm, leading to the festival being known as "the punishment of the dogs."

In the Xinjiang Province of China, geese are used to guard a police station and other buildings. A report by *The Telegraph* newspaper in 2013 tells how the police chief, Mr. Zhang, welcomes the helpful stewardship of their geese. "In some ways, they are more useful than dogs," he told reporter Tom Phillips. "A household normally keeps one dog, [but] an intruder can throw a drugged bun to kill the dog. Geese are normally kept in groups, and they have poor eyesight at night, making it very difficult for intruders to [poison them]."

Geese were used to protect a Dumbarton, Scotland, warehouse where the famous Ballantine's whiskies were aged. These geese served from 1959 until 2012, when modern CCTV cameras were added. Known as the

"Scotch Watch," the fierce white geese were the subject of numerous TV documentaries and magazine articles.

Utilizing Guardian Geese

Even if you aren't using your geese to guard, their skill as alarm bells is amazing. Anything out of place in the barnyard or approaching unexpectedly will be greeted with rowdy honks. If you are familiar with your flock and spend enough time in their presence, you will be able to distinguish their alarm honks from mating calls and general chattiness. This talent can be invaluable, giving you extra minutes to get to your chickens before a fox attacks or just letting you know that the mail has arrived. They are vigilant about possible aerial predators such as hawks, and will notice a potential threat long before you or your chickens do.

Geese make intimidating watchdogs for your property.

Geese employed as alarms need to be kept in a place with a good vantage point over the rest of the farm, and should be enclosed within their space. Because of their wandering nature, it's not uncommon for unfenced geese to get distracted from the job of watching out for strangers.

Nighttime, however, and the goose's advantage as a guard animal diminishes: they are virtually blind at night. They will still sound a warning if they notice any movement, but they won't react until much later than they usually would in daylight. Also, at night, they are somewhat clumsy in identifying what the threat actually is. Note, as well, that geese will try to expand their territory if not fenced in, and may end up attempting to guard the neighbor's lawn from the neighbor. If you live in a more suburban area, keep your geese fenced away from areas they should not approach and always make sure they are safely enclosed at night.

Using geese for guarding does have its risks. The first one is the safety of your geese, fearless attackers who sometimes take on more than they can handle. While a goose will effectively scare off any predator smaller than a cat, and can even chase away a timid dog or human, against a coyote or larger predator, they will lose the battle. The best way to avoid problems is to be smart about how you are using your guard geese.

Geese make great protectors of smaller flocks, such as chickens, ducks, and quail. They can also protect a building or your personal possessions. While they are excellent at sounding the alarm, be aware that they might succeed in chasing off only a few intruders. Thieves and intruders are more sophisticated than they were in Rome in 390 BC.

To Protect Your Flock

To protect smaller poultry, it is important to use only a single guard goose or possibly a pair, but no more than that. One goose per flock of chickens or other poultry helps to convince the goose it is related to the chickens, and therefore needs to defend and protect them. To further help the goose bond with its congregation, raise your guard goose from a gosling with the chicks. Similar to how geese imprint on humans if raised with lots of handling, goslings nurtured with chicks and not other geese will imprint on them and have a stronger sense of flock with them, promoting better

defense of the hens. They will also get along better with your chickens this way, whereas geese raised separately may try to peck at or chase smaller birds.

A goose used for the guarding and alarm work of a chicken flock may not be as interested in, or bonded to, people as they are to their farmyard companions. They will still recognize their owners, and will limit defensive attacks and aggressive behavior only to what they perceive as threats to their friends. Guard geese will share some of the chickens' scratch and water access, but they are happy to graze the pastures, making them a very affordable option to help in maintaining the safety of your hens.

Most geese used as alarms are not specifically kept for that purpose. Any goose on the farm can be a protective single guard, thanks to their keen eyesight, clamorous voices, and intimidating behavior. However, if you want geese specifically because of their ability to guard, it is a good

A young African goose watches over a flock of hens.

idea not to imprint with them as closely as you might with pet. If you are not guarding other poultry, but want to keep your property safeguarded, try keeping a larger flock for intimidation purposes. While a single goose is no match for a large dog or coyote, a full flock of them may be enough to scare off such predators. And geese can certainly be effective at intimidating people, especially when individuals harbor pre-existing ideas or phobias regarding geese.

The geese on our farm aren't employed as guards, but they are still successful at that task. As our flock has grown, we've noticed a decrease in attacks on our chickens, and we never have a neighbor stop by without hearing the geese announce the call long before the visitor has reached the door. Because our geese have imprinted on humans and are accustomed to being physically handled, they don't attack strangers; they simply alert us to their presence. Occasionally, they will run towards a person, but they always stop short of making a full charge.

The great thing about guardian and alarm geese is they don't need any training to do their job. Protecting is a natural instinct for them; they will take to it like ducks to water. This can be a handy advantage to backyard geese, and many farmers see this as a primary reason to invest in these feathered sentries. A goose for every flock of chickens can ensure that everyone, geese and hens alike, is safe and happy.

Chapter Eight:
Weed Control

*G*EESE HAVE A BOTTOMLESS APPETITE for fresh green grass. This constant hunger can easily be utilized by the resourceful farmer. Whether you have a small hobby farm or are keeping large commercial crops, the appetite of geese can be an advantage for your farm.

Geese don't enjoy broad-leaved plants as much as sweet blades of green grass. This means that geese can be successfully and reliably employed to help maintain farm and garden areas ranging from a small onion patch to more spacious ones dedicated to nursery stock or orchards. Weeding paths around small trees or shrubs, a large flock of geese can remove the need for extra human labor. With a little bit of know-how, you can keep your birds from gobbling up the stuff that you don't want, instead of destroying the sections that you do. Understanding how to manage weeder geese will help establish a good working relationship with them.

Take into account that hungry geese might eat up the very crops you are trying

Geese will enjoy most leafy greens.

79

to grow. Geese can be used for weeding a wide variety of crops, but keep them away from certain veggies that they will chow down indiscriminately. They'll enjoy the lettuce, tomatoes, or pansies just as much as you do. Geese should not be used for weeding a small kitchen garden or a flower bed. If you keep geese and have a garden, it is actually a good idea to surround it with a small solid fence to keep them from enjoying your hard work. A fence need not be more than 2 to 3 feet tall, with no large gaps a goose can fit through. So-called no-climb or stock fencing works well, but a small picket fence can make the fencing less obtrusive and even add some charm.

A flock keeping a field trim.

Orchards and vineyards are great places to use geese for weeding. They will love the job of maintaining clean paths around tree trunks or the climbing tendrils of vineyards, and they can be equally effective when weeding other fruit bearing vines such as strawberries.

Among the best places to use geese for weeding are:

- Apple orchards
- Vineyards: grapes, hops, or other vines
- Strawberry patches
- Raspberry and blackberry patches
- Small lawn areas
- Tobacco fields
- Potato fields

You can start using your geese for weeding when they are still downy goslings.

- Garlic or onion patches
- Many herbs, depending on the goose
- Evergreen trees, including Christmas tree farms

Geese can develop an appetite for certain weeds more than others if you feed them regularly as treats when they are young. Unfortunately, training them *not* to appreciate plants is much more difficult, but geese are picky creatures, and if you know your birds well, you can perceive which ones they will seek out or pass up. You'll want to watch your young geese initially and learn to recognize their preferences.

At our small farm in Maine, we use geese to keep our lawn trimmed and intend to use them more extensively for weeding as we plant more crops. A flock of ten geese can keep about an acre manageably trim, though not as perfect as a manicured suburban front lawn. In a dry summer, with their help, we don't need to mow from the beginning of August until frost.

Why would you use geese to keep your crops weeded? Especially around vines and crawling plants, their long, thin necks can reach in among the growth and fruits and pick out weeds that human eyes might otherwise overlook. Weeder geese are obviously far less toxic than chemical sprays or herbicidal applications, and often more effective. The Maine Organic Farmers and Gardeners Association describes geese as "the closest foragers known," applauding their studious consumption of weeds.

Geese used for weeding are remarkably easy keepers. Apart from a handful of grain in the evenings, weeder geese do not need access to food beyond the greenery they're "employed" to eat. They will maintain a healthy weight, but they will be just hungry enough to continue foraging constantly. This lack of a need for constant grain makes them extremely inexpensive to feed, and all they need in addition is freshwater for eating, drinking, and bathing. Harvest by-products from other areas of your garden, such as carrot and beet tops and extra salad greens, to further supplement their grain diet and reduce feeding costs.

Chemical weed killers are dangerous to your crops as well as the environment, often expensive and unreliable. But geese actually improve the quality of your soil as they work. Your plants and soil will flourish with the enrichment of their fresh goose droppings, a natural fertilizer rich in nitrogen.

The use of geese for weeding purposes dates back several decades, if not much longer. In March 1986, the *Los Angeles Times* (articles.latimes.com/1986-03-11/news/mn-3148_1_weeder-geese) ran an article on a farmer in Riverdale, who kept a lucrative business renting out "weeder geese." Fernando Alves was charging $4.25 per goose to various crop farmers in Southern California. At the time, he was renting over 20,000 of his White Chinese geese a year.

Commercial use of geese for weeding is relatively uncommon now, as massive monoculture farms have little time for anything but chemical sprays. Yet many smaller farms still take advantage of geese as an environmentally friendly weed control option. Hoch Orchard and Gardens (hochorchard.com) in La Crescent, Minnesota, use a flock of geese for weeding their strawberry beds. They claim that they have reduced the need to hand weed by as much as 90%. The birds also provide extra income to the farm from marketing them as orchard-raised geese for meat. Geese are used on Christmas tree farms and vineyards around the world, helping farmers keep their plants healthy and their costs down.

To properly care for weeder geese, you will still want to provide nighttime shelter to keep them safe from predators. Monitor their progress daily and move them to new areas of pasture regularly. Left on a small area for too long, they can eliminate growth too completely. On the other hand, if geese are moved to new areas too soon, then weeds might quickly reappear. The farmer needs to learn how to manage the geese, according to weed growth and the type of crop.

Even if you don't use your geese for weeding, you can still take advantage of their nitrogen-rich droppings. You can spread fresh compost from their living quarters on your flower or vegetable beds and watch your plants flourish. Dried goose droppings contain about a 2:4:2 nitrogen-phosphorus-potassium ratio, so it can be strong for your garden if applied directly, but used in combination with other compost and debris, it will work wonders. You can also use goose droppings to create "manure tea" in place of chemical fertilizer.

The right breed of goose makes a big difference in their effectiveness as weeders. The White Chinese is a favorite, thanks to its high energy level and extra-long and slim neck. The larger breeds, while ideal for meat birds, are not as effective as weeders because they aren't as interested in foraging for their food, and also might crush vegetation due to their heavy frames.

Chapter Nine:

Goose Recipes

$\mathcal{M}$OST OF US ARE FAMILIAR WITH THE ENGLISH CHRISTMAS CAROL lyric "Christmas is coming, the goose is getting fat." Indeed, there are many families whose holiday traditions are not complete without a golden brown roasted goose on the table. If you've never tasted it, then the best introduction for you might be to get an organic goose from a local farmers market and try it for yourself. You may decide that roast goose is the reason to raise these birds on your farm. Then again, if you're undecided about goose fare but would like to raise geese, you can extend your culinary experiences with them by cooking with their eggs. Goose eggs can be substituted in any recipe that calls for chicken eggs, at a ratio of one goose egg for three chicken eggs. You'll appreciate the difference in richness and smooth flavor.

Whether it's roast goose, omelets, or even goose egg pasta, you'll find that cooking with goose products is just one more benefit of having a gaggle of them around your home.

Goose eggs are as big as three chicken eggs.

Roast Goose

Roast goose is a classic holiday dinner, and it is not difficult to turn the full carcass of a mature goose into a mouth-watering alternative to turkey or ham.

Ingredients

1 whole goose, 10–15 pounds
1 tablespoon fresh thyme
1 tablespoon fresh rosemary
1 onion, cut in half
3 carrots, cut in half
1 bay leaf
Salt and black pepper

Directions

Preheat oven to 400°F.

Remove any giblets or pads of fat from inside the bird. Using a sharp knife, score across the breast and legs in a crisscross pattern.

Rub the goose inside and out with the rosemary and thyme, and ample salt and pepper. Place in a large roasting tray and arrange onion, carrots, and bay leaf around the goose.

Place in the preheated oven. After 10 minutes, reduce the temperature to 350°F.

Every 30 minutes, baste the bird with the pan juices and remove any fat drippings from the pan. The fat can be saved for later cooking experiences.

If your goose is browning too quickly, it can be covered with aluminum foil to protect the skin.

Depending on its weight, the goose should take about 3 hours to cook through. Check its readiness using a meat thermometer, which should read 180°F when inserted into the breast away from the bone.

Once done, remove the goose from the oven and let sit for 10 minutes before carving.

Goose and Kraut

A Germanic take on the roast goose, this is an especially festive recipe, and the tart sauerkraut perfectly complements the rich meat.

Ingredients

1 whole goose, 10–15 pounds
1 pound of sauerkraut
1 teaspoon sugar
1 cup boiling water
1 lemon
Salt and black pepper

Directions

Preheat oven to 350°F.

Half your lemon and use the open faces to rub the goose carcass inside and out.

Salt and pepper your bird liberally.

Combine the sugar and sauerkraut in a mixing bowl, along with an additional dash of salt and pepper.

Stuff the carcass with the sauerkraut mixture. Set in a large roasting pan, breast up.

Pour the boiling water into the pan.

Cover the bird with foil and roast in the oven for 1 hour.

Uncover the bird and remove any fat drippings from the pan, then return it to the oven.

Depending on the size of the bird, it should be done within another 1 hour or 2. Test its readiness with a meat thermometer, which should read 180°F when inserted in the breast away from the bone.

Once done, remove from the oven and serve with the sauerkraut stuffing and other roast vegetables.

Goose Breast

One of the easiest ways to enjoy goose meat is to use the breast fillets. This is also a less expensive option if you're purchasing your goose, and it's just the right amount of meat for one or two.

If you are taking your geese to a butcher to be slaughtered, they will cut out the breast for a fee. If harvesting your own birds, you can do this yourself, a very common practice among goose and duck hunters. After plucking the bird, make an incision through the skin of the breast so that the meat is exposed. Peel the skin aside, so that all of the breast meat is exposed. With a sharp, preferably flexible, knife, cut along the side of the breastbone all the way down the ribcage, and then follow the ribcage closely with your knife to cut out the breast. Cut along the collarbone and then carefully around the wing bone to completely free the breast meat.

This simple method, which hunters have employed for centuries, should leave you with two clean, plump fillets.

Ingredients

2 goose breasts, trimmed of fat
Salt and black pepper to taste
2 tablespoons butter

Directions

Season each breast on both sides with a liberal dash of salt and pepper. You can also add any other seasonings that you prefer; a personal favorite is garlic powder.

Heat the butter in a large skillet over medium heat.

Once the butter is melted, add the breasts to the pan. Cook on one side for 4 minutes, then turn them over and cook for 4 minutes on the other side.

Serve rare or medium rare with roast vegetables.

Goose breasts can also be oven roasted. Preheat your oven to 350°F and prepare the breasts in the same manner. They will cook for about 30 minutes for a medium rare fillet.

Foie Gras, Goose Pâté

One of the most famous goose dishes in the world is foie gras. While the traditional method for raising foie gras geese is exceptionally cruel, goose liver pâté can be made with the liver from any healthy farm goose that has been humanely pasture raised.

It will take 5 or 6 goose livers to make a small plate of pâté. This is both wasteful and impractical if that is all you are raising your geese for. But if you are slaughtering the birds for meat anyway, the livers often go unused unless you're resourceful enough to try this recipe.

Ingredients

1 pound of fresh goose liver, trimmed of membranes and fat
1 stick (8 tablespoons) unsalted butter
1 tablespoon brandy or port
1 garlic clove, minced
Dash of salt and black pepper
2 shallots, chopped
1 teaspoon fresh thyme leaves, chopped
¼ cup heavy cream

Directions

Rinse and pat the livers dry. Place them in a bowl with the brandy and leave refrigerated for at least 2 hours.

Melt the butter in a skillet over medium heat.

Place the livers in the skillet and add the shallots, garlic, and thyme. Season with salt and pepper.

Set the soaking brandy aside — you will be using this later.

Cook over medium heat for 3 minutes.

Transfer the livers, shallots, garlic, and thyme to a blender. Add the soaking brandy. Blend until liquified.

Add the cream, and pulse until combined.

Season the mixture with salt and pepper and spoon into a small bowl or terrine. Put the mixture through a strainer if a finer texture of pâté is desired. Refrigerate 2 hours before serving.

One-Egg Omelet

One of my favorite ways to utilize the massive eggs geese lay is in making omelets. A single goose egg, which is equivalent to three chicken eggs, is perfect for one or two servings, depending on your appetite.

Ingredients

1 goose egg
1 teaspoon vegetable oil
Dash of salt and black pepper
1 tablespoon dried thyme
2 tablespoons shredded Cheddar cheese

Directions

Crack the goose egg into a large bowl with salt and pepper. Whisk thoroughly.

Heat vegetable oil in a large skillet over medium heat.

Pour the whisked egg into the skillet. Cook until it starts to thicken and turn brown.

Sprinkle the cheese and thyme over the top of the omelet, and carefully fold one half of the omelet over the other.

Transfer to a serving dish and enjoy.

Goose Egg Pasta

In Italy, legendary for its pasta, goose eggs are prized for their thick yellow yolks. Chefs throughout the culinarily renowned country talk of the tenderness of goose egg pasta, and if you're going to try making your own noodles, goose eggs are the way to go.

Ingredients

2 goose eggs
4 cups all purpose flour
1½ tablespoons water
½ teaspoon salt

Directions

Whisk together the flour and salt in a large mixing bowl.

Create a large cavity or "bowl" in the center of the flour mixture. Crack the eggs into the cavity.

Whisk the eggs inside their cavity, not yet combining them with the surrounding flour.

Gradually begin to whisk the flour into the eggs, pulling from the walls of your flour "bowl." Combine slowly, do not rush the process.

Once enough flour has been combined into the eggs, the mixture will form a soft dough.

Turn the dough onto a clean, freshly floured countertop.

Gently knead the dough, folding it into itself carefully with the heels of your palms. As you knead, you will feel the dough firm up.

You'll know that the dough is ready when it forms a tight ball and few air bubbles show in the mix if it is cut with a sharp knife.

Rest the dough in a clean, dry mixing bowl covered with plastic wrap for 30 minutes.

Prepare a baking sheet for the dough by sprinkling it liberally with flour. Divide your rested dough into 4 equal sections and place them evenly on the baking sheet. Cover with a clean dish towel.

Take one disc of dough and flatten it into a thick cookie between your hands.

Using a pasta machine set to the thickest setting, gently feed the dough through, and as you catch it with your hands, fold into thirds as you would a letter. Feed it through 1 or 2 more times, repeating the folding process, at the thickest setting.

Set your pasta machine at the next thinnest setting and roll the pasta through 2 or 3 times. Do not fold. Continue working your way down the settings until you have reached your desired thickness of pasta. Do not skip settings.

Once your pasta is at the desired thickness, lay it out on a smooth, lightly floured surface. With a sharp knife, slice into your preferred length and thickness of noodle. For rounded noodles, roll them into your preferred shape.

Toss the cut noodles with flour and place them in a bowl covered with a clean dish towel.

You can now either cook your fresh pasta immediately, freeze your noodles, or set them to dry and then store them.

Rendering Goose Fat

When you cook at all with goose, you'll have some fat drippings to set aside and use to cook vegetables, fish, and much more. If you really want to make the most of your goose eating experience, you can also render the fat trimmed from the carcass. This is perfect drizzled over roast potatoes, surrounding a pan-fried sea bass, or for making Yorkshire pudding. Use in the place of bacon fat or olive oil in many recipes.

Ingredients

Fat trimmings from a fresh goose carcass

Directions

Cut the fat into small pieces and place them in a large skillet over low heat.

The chunks of white fat will turn brown and slowly shrink, leaving behind a slick, shining liquid. The time varies depending on how much fat your bird had, but you are done when only small brown nuggets remain at the bottom of the skillet.

Carefully strain the liquified fat into prepared Mason jars. Cheesecloth is best for straining.

You can enjoy the remaining "cracklings" of fat left behind, and store your rendered fat in a cool, dry place for up to 12 months.

Goose Egg Desserts

Goose eggs make delicious pastries. They will thicken a batter quickly and leave it a rich orange. The flavor of cakes and pies made with goose eggs is deeper. You can substitute a single goose egg for three chicken eggs in any dessert recipe, and they are particularly delicious when used in brownie mixes. The use of a goose egg for dessert can turn an ordinary custard or pudding mix into something truly exceptional.

A bantam hen's egg compared to a goose egg.

Goose Egg Custard

Goose eggs make a rich and creamy custard, a delicious after dinner treat. Custards are favorite way to use up extra goose eggs, and a batch of custard never lasts long.

Ingredients

2 goose eggs
4 cups of milk
½ cup of raw honey
1 tablespoon of vanilla
Pinch of salt

Directions

Start by preheating your oven to 325 degrees. Place a pan or baking sheet in the oven with 1" deep water in it to heat.

Combine the eggs, honey, salt and vanilla in a large bowl and whisk together.

Heat the milk in a saucepan and then slowly stir the milk into the egg mixture. Do not mix them too fast, as the milk will cook the eggs.

Stir constantly until all of the ingredients are combined.

Pour the mixture into ramekins and place the ramekins on the baking sheet with the hot water in your oven.

Bake for 40–50 minutes.

Custard can be enjoyed hot or cold, and it is most delicious with a little bit of cinnamon sprinkled over the dish.

Goose Egg Meringues

Meringues are beautiful and tasty, and a goose egg meringue is sure to impress guests and please the taste buds.

Ingredients

1 goose egg white
¾ cups of sugar
pinch of salt
⅛ teaspoon cream of tar tar
1 teaspoon vanilla extract

Directions

Preheat your oven to 250 degrees. Prepare a baking sheet with parchment paper.

Beat the egg white together with the salt and cream of tartar. Mix until peaks start to form. You can mix by hand, but using an electric mixer or food processor will make this task much easier.

Slowly add in the sugar and continue to mix, maintaining stiff peaks.

Add in the vanilla and mix until fully blended.

You can add a few drops of food coloring for a splash at this point.

You can use a piping bag to lay out shapes and forms with the mixture on your baking sheet, or simply dollop spoonfuls of the meringue mixture onto the sheet.

Bake for 20-30 minutes and allow to cool fully before serving.

Goose Eggs for Breakfast

They're huge, but that doesn't mean you cannot make ordinary fried eggs with goose eggs. You can also hard boil or scramble them. You may notice a slightly gamier flavor and will see that the yolk-to-white ratio is much greater. A hard boiled goose egg sliced open will reveal how much of the inside is deep-orange yolk.

Because they're bigger, goose eggs will take longer to cook in a skillet or boiling water. Allow 10 to 12 minutes when hard boiling goose eggs, and give yourself a few extra minutes on any stove-top recipe.

Goose Egg Waffles

If you have a waffle iron, you know how perfect waffles are for a meal or a snack. A goose egg can further transform your waffles into fluffier, richer meals.

Ingredients

2 cups of all purpose flour
1 tablespoon of baking powder
3 teaspoons of sugar
1 goose egg, separated
1½ cups of milk
6 tablespoons of butter, melted.

Directions

Prepare your waffle cooking surface to begin.

In a large bowl, sift together all of the dry ingredients.

In a separate bowl, whisk together the egg yolk, milk, and melted butter.

Slowly mix the egg mixture into the flour mixture.

In a separate bowl, mix the egg whites together until peaks start to form.

Gently fold the egg whites into your waffle mixture.

Ladle your mix onto your waffle iron ¼ to ½ cups at a time, and cook each waffle for 4-5 minutes.

Serve with your favorite waffle topping.

Goose Egg French Toast

My go-to weekend meal is French toast, and I love it made with goose eggs. My favorite way to mix it is with a little bit of extra vanilla, and served with warm syrup and butter.

Ingredients

1 goose egg
¼ cup of milk
1 teaspoon cinnamon
1 teaspoon vanilla extract
8 slices of white bread

Directions

Combine the egg, milk, vanilla and cinnamon in a large bowl.

Dip your bread, slice by slice, into the mixture and make sure that each side of the bread is thoroughly coated.

Fry the bread slices on a warm skillet until golden brown.

Chapter Ten:

The Joy of Keeping Geese

I'VE KEPT POULTRY ALMOST MY ENTIRE LIFE. Chickens, ducks, guinea fowl, even a parrot. But no bird has brought me as much delight as my geese. From the moment I scooped Lady Goose out of her little cardboard shipping box and stroked her gray, downy fluff, I was hooked. It wasn't just the adorable peeping, the soft down, or her miniature black beak. Every goose I've ever owned has had a glimmer in their eye, which I personify as intelligence and love.

Lady Goose, our first goose.

When a goose imprints on you, it doesn't feel like they're making a random choice between the various blurs of people who bring them food. It feels like they've purposefully selected you, trust you, and want to be around you. A goose might bond with one person in a couple and not the other, despite equal attention from both people.

Not all geese imprint, especially not when being raised with many other goslings or when they're introduced to a flock of adult geese while still young. When raising a trio of goslings a few years ago, I noticed that two of them had no real interest in me or people, and were friendly but cautious. The third, a Roman Tufted gosling we named Pete, imprinted wholeheartedly upon me. Pete was never more than a few feet away from me all summer long, and he would cuddle up under my shoulder so that I could rest my head on his feathered back. He tolerated being picked up and snuggled, but he adored nibbling at anything and everything I was wearing, even my sunglasses. Not once in that intimate setting did

There's nothing like goose love.

he show any of the fearful hissing and honking behavior that is commonly associated with geese.

When I was a small child, not more than eight or ten, I remember going with my father to a nearby farm to pet the horses. In the farmyard was a gander, a big Toulouse with a pendulous jowl, and his friend, a slender white goose. I was terrified of them. My dad pretended not to notice as the big goose chewed around the edges of his shoe as we stood with the horses, and I held onto his hand for dear life. Such is the reputation of the fearsome goose, that even a subdued display such as this could strike fear into the heart of a young girl.

If I'd known then what I know now, I'd realize that standing your ground and not showing fear would more than likely cause this big gander to back off. I'd also know that his nibbling was barely aggressive, a testing of his limits sparked more by curiosity than rage. Perhaps my love of geese started at this moment, a decision to confront fear and befriend the behemoth.

Rather like the maiden taming the fire-breathing dragon, there is a satisfying challenge in befriending a goose. And whether or not you are using your geese as guard animals, there's something special about a bird that will hiss at strangers to protect you. It is not why we keep geese, but it is among the reasons that we enjoy their presence on the farm so much.

It is a myth that geese are monogamous birds. Wild Canadian geese may be, but farmyard ganders take three or four mates in a year. However, their emotional attachment to their little harem is remarkable. A goose who has lost its gander will honk sorrowfully for weeks or months, overcome with loneliness. Even if a gander keeps a few geese as mates, he'll notice if one is missing and call for her.

Watching geese is end-lessly entertaining.

It is very clear in a large flock where the family divisions are. Geese largely keep to themselves within a family unit, the gander hissing off any geese that are not either his mates or offspring. While their arrangement may not be as romantic as rumor has it, it is still inspiring to see how deeply these birds care for each other.

The bond you can form with your geese is unusual in the animal kingdom, especially among farm animals. As if in direct contrast to their fiery reputations, a bonded goose is one of the sweetest and most endearing creatures you can keep in your backyard.

While the bond of a grown goose is an unbreakable adoration, the connection you can make with goslings after only a few days of caring for

Pete offers some affectionate kisses.

them is even more special. The hardest of souls will soften in view of a troupe of goslings urgently calling out and running, tiny stubs of wings spread, after their adoptive "mama." I think it is these moments that change many people from being farmers who happen to be raising geese into dedicated and loving goose farmers.

I have had special bonds with many of my geese, especially Lady Goose and Pete as I have described here. But even the geese who never paid me much attention, the ones who never imprinted, have a special place in my heart.

If you're a farmer, or even keep a small backyard homestead, you've probably personified your animals a few times. Geese lend themselves to these amusing activities and storylines beautifully. In fact, I could sit and watch a flock of geese all day and smile at their interactions and human-like emotions. The flock hierarchy isn't as vicious as with chickens, and it seems to include more gossip than even clucking hens.

In our flock, there's the diminutive Sebastopol gander who has fallen in love with the giant Dewlap Toulouse goose. He spends his days walking in circles around her, plucking blades of grass for her and hissing off anyone who tries to approach. She, meanwhile, looks at him fondly but carries on eating as if it didn't really matter if he was there or not.

We have the slightly neurotic African goose, who every spring cautiously approaches any chick, duckling, or gosling within earshot to coo and mutter over them as if she's saying, "Are you my babies?" And let's not forget Pete, our Roman gladiator goose, who quickly established himself as the head gander and struts his stuff off in the distance, away from the rest of the flock with his harem of three African geese.

Like many farmers, when I first got geese, I was not expecting them to form emotional bonds with me, nor was I looking for all the personalities that would brighten our farmyard. These came as unexpected bonuses to a plan that had centered on guarding our chicken flock and enjoying giant eggs. Now that I have been keeping geese for many years, my reasons for enjoying them and their continued presence on our farm have shifted. Yes, they offer us delicious eggs, and yes, they keep many would-be predators at bay. But sitting sipping my morning coffee and enjoying the sight of a flock chatting away about the day's activities makes keeping geese a pleasure. And that pleasure doubles when someone like Lady Goose decides to wander over and, out of politeness, include me in the daily gossip.

Acknowledgments

*T*HIS BOOK COULD NOT HAVE HAPPENED without the support and encouragement of some wonderful people.

As a child my first introduction to geese was with my friend Anna Nellis Smith, who had a pair of Toulouse geese on her family farm. Her confidence with them was a challenge to me, and soon I was just as fond of the "goosers" as she was.

Once an adult, I did not consider adding geese to our farm until the request of my partner's son, Wallace Jackson. Wallace's inspired idea led to our first geese (Mr. and Mrs. Goose) who first introduced me to the wonders of imprinting and the delights of keeping these birds.

I could not have written this book without the assistance of my mother, Karyn Lie-Nielsen, who instilled in me a love of language and words from a young age. As I worked on this, she was an invaluable source of feedback and motivation. Without her, I never would have considered myself a "writer."

My partner, Patrick Jackson, has been extremely patient and supportive throughout this process. For years he's been telling me to follow my dreams and to find what moves me, and our transition to the country and into keeping geese and other animals has been everything I ever dreamed of. I know that his never-tiring drive will continue to bring us new adventures.

My father, Thomas Lie-Nielsen, was also encouraging and supportive throughout the process of writing this book, and he along with all of my co-workers were very patient as I took the time needed to complete this project.

My thanks go out to Lisa Steele, author of *Fresh Eggs Daily* and other titles, for her support and help in finding a home for this book. Lisa was an inspiration to me before we met, and continues to be an encouraging friend.

Marta Madden was able to come out to the farm and take some beautiful pictures at a moment's notice when this book deal was first approved. She also has been a constant supporter of our growing farm.

Kate St Cyr, who has been an enthusiastic supporter of mine and someone to bounce all sorts of homesteading ideas off of, I must thank immensely for her beautiful photography and continued support.

To everyone who was patient with me throughout this process, helped with selecting pictures and phrases, and encouraged my love of geese, my sincere thanks. I never imagined at the beginning of our farm, when we were caring for a couple of chickens and planting our first tomatoes, that it would grow into this. Thank you for understanding, for listening, and for believing in me as I took leaps of faith that have paid off delightfully.

And to everyone who buys this book or adds geese to their farms and backyards, thank you for helping to dispel their negative reputations and for being among the growing number of people who know that there's no better fowl friend than a goose.

Resources

Gosling Hatcheries

Metzer Farms
Gonzales, California
1-800-424-7755
www.metzerfarms.com

Holderread Waterfowl & Farm Preservation Center
Corvallis, Oregon
541-929-5338
www.holderreadfarm.com

Sand Hill Preservation Society
Calamus, Iowa
563-246-2299
www.sandhillpreservation.com

Poultry Supplies and Information

Tractor Supply
www.tractorsupply.com

Backyard Chickens Forum
www.backyardchickens.com

Goose Breed Information

The Livestock Conservancy
Pittsboro, North Carolina

919-542-5704

www.livestockconservancy.org

American Poultry Society
Burgettstown, Pennsylvania
724-729-3459
www.amerpoultryassn.com

Recommended Reading

Holderread, Dave, *The Book of Geese*
Hen House Publishing Co., 1993

Backyard Poultry Magazine
1-800-551-5691
www.countrysidenetwork.com

GRIT magazine
1-866-803-7096
www.grit.com

Index

About the Author

$\mathcal{K}$IRSTEN LIE-NIELSEN grew up on a farm and has been raising geese and enjoying the quirky personalities and practicality of these beautiful birds for most of her life. Always intrigued by self-sufficiency and working with her hands, Kirsten and her partner are restoring a 200-year old farm in Liberty, Maine, where they raise animals and grow vegetables and native medicinal herbs. Kirsten writes about her experiences and the lessons of life on the farm for publications such as *Grit* and *Mother Earth News*, *Backyard Poultry*, and Hobbyfarms.com. She also blogs about the good life at hostilevalleyliving.com.

A Note about the Publisher

NEW SOCIETY PUBLISHERS is an activist, solutions-oriented publisher focused on publishing books for a world of change. Our books offer tips, tools, and insights from leading experts in sustainable building, homesteading, climate change, environment, conscientious commerce, renewable energy, and more — positive solutions for troubled times.

We're proud to hold to the highest environmental and social standards of any publisher in North America. This is why some of our books might cost a little more. We think it's worth it!

- We print all our books in North America, never overseas
- All our books are printed on **100% post-consumer recycled paper,** processed chlorine free, with low-VOC vegetable-based inks (since 2002)
- Our corporate structure is an innovative employee shareholder agreement, so we're one-third employee-owned (since 2015)
- We're carbon-neutral (since 2006)
- We're certified as a B Corporation (since 2016)

At New Society Publishers, we care deeply about *what* we publish — but also about how we do business.

New Society Publishers
ENVIRONMENTAL BENEFITS STATEMENT

For every 5,000 books printed, New Society saves the following resources:[1]

16	Trees
1,483	Pounds of Solid Waste
1,631	Gallons of Water
2,128	Kilowatt Hours of Electricity
2,695	Pounds of Greenhouse Gases
12	Pounds of HAPs, VOCs, and AOX Combined
4	Cubic Yards of Landfill Space

[1]Environmental benefits are calculated based on research done by the Environmental Defense Fund and other members of the Paper Task Force who study the environmental impacts of the paper industry.

MIX
Paper from
responsible sources
FSC® C016245

new society
PUBLISHERS
www.newsociety.com